BEST RADIO PLAYS OF 1983

The Giles Cooper Award Winners

Wally K. Daly: Time Slip
Shirley Gee: Never In My Lifetime
Gerry Jones: The Angels They Grow Lonely
Steve May: No Exceptions
Martyn Read: Scouting for Boys

METHUEN/BBC PUBLICATIONS

First published in Great Britain in 1984 by Methuen London Ltd,
11 New Fetter Lane, London EC4P 4EE and in the USA by
Methuen Inc, 733 Third Avenue, New York, NY 10017, and
BBC Publications, 35 Marylebone High Street, London W1M 4AA.

Typeset by Words & Pictures Ltd, London SE19
Printed in Great Britain by
Redwood Burn Ltd, Trowbridge, Wilts

British Library Cataloguing in Publication Data

Best radio plays of . . . (Methuen modern plays)
   1983
   1. Radio plays, English————Periodicals
   822'.02'08   PR159.R33

ISBN 0-413-55220-9

CAUTION
These plays are fully protected by copyright. Any enquiries concerning
the rights for professional or amateur stage production, broadcasting,
readings etc., should be made to the authors' respective agents, as
follows:
Wally K. Daly: Bryan Drew Ltd, 81 Shaftesbury Avenue, London WC1
Shirley Gee: David Higham Associates, 5-8 Lower John Street,
   London W1R 4HA
Gerry Jones: 12 Grange Avenue, Strawberry Hill, Twickenham,
   Middlesex
Steve May: 10 Trinity Place, Merchants Road, Hotwells, Bristol 8
Martyn Read: Harvey Unna and Stephen Durbridge Ltd.,
   34 Pottery Lane, London W11

CONTENTS

PREFACE

1983 was an eventful, not to say a fateful year for radio drama both in Britain and abroad. There were doubts and debates and uncertainties; there were pressures of planning and economy; there were welcome revivals and bursts of creativity; and there was at least one demise — though perhaps not a total one.

The doubts and difficulties have long been experienced most keenly in the United States, where radio drama has continued to exist at all only in small pockets, sustained by individual professional enthusiasm against all the odds. It would be an exaggeration to claim that these notable, largely personal achievements have had much effect on the world of the creative arts in America, let alone on the general public, but it's a curious fact that good radio plays have continued to spring up from time to time, like bright-flowering cacti in a desert, in both commercial and public service stations. In a way it says much for the unique and enduring qualities of the form that it can survive however sporadically in a place where scarcely a memory — let alone any professional tradition — of its origin and development remains.

Sadly, the year witnessed a major crisis in the organisation to which one might have looked for a more permanent revival and real nourishment of the radio play. National Public Radio, the association of local public service stations with a number of production centres for specialised, expensive programming, had only recently established a regular service of drama — ranging from popular serials such as the radio version of *Star Wars* to the more experimental single programmes of *Earplay*, which embraced both plays and dramatic features. Plagued by an inability to augment federal funds with corporate finances for drama projects, in 1983 NPR also suffered major cuts in both direct and indirect government funding. As a result, although the organisation itself just escaped from total collapse, it did so at the expense of losing many of its creative staff, together with the drama and arts programming for which they were responsible. It looked like the end of the road for the last gallant defenders of the form.

And yet, as it were from the very ruins of the Alamo, (in fact from an unlikely combination of a writer in Fort Worth, Texas, and a producer in Madison, Wisconsin), five radio plays of stature have just completed production and will reach the air soon in the United States and in Great Britain. More may follow. It is indeed slender evidence of indestructibility but nature has provided even more curious and triumphant examples.

Nearer home, in Europe, there continue to be discussions in many broadcasting organisations, if not about the very survival of the radio play, at least about its form and place in the context of broadcasting as a whole. The biennial conference of radio drama experts under the auspices of the European Broadcasting Union regularly debates 'the concept of the short radio play', managing to sweep aside as frivolous the notion that a short radio play is simply of lesser duration than a long one, and earnestly pursuing the idea that the short radio play is somehow a new art-form for our impatient times. The question will undoubtedly be raised yet again in June of this year at the conference in Geneva and, sadly, I believe that there may be more evidence on offer that the idea is being put into practice.

It's not that I'm against short plays; on the contrary, the BBC has over the years broadcast many of them. I think with particular affection of the late Alan Gosling's brilliant duologue *Little Black Hole*, which in the space of just over eleven minutes exposed the missed opportunities of a married lifetime and made the listener care, desperately, for the elderly couple who were at its centre. More, it exploited with great delicacy the advantage of not being able to see the action reserving till the last minute the revelation that one of the protagonists was actually attending the burial of the other. The conversation took place, simply and naturally, after death — and was cut short only by the coffin descending into the ground.

No, it's not the length of the play that concerns me but the idea, which so often underlies the argument, that it is the listerner's reduced attention span for which we have to cater; or worse, that drama has to be 'packaged', polythene-wrapped and attractively (or even deceptively) labelled within a magazine format so that it slips unnoticed into the shopping basket of the mind. And that, of course, is the very antithesis of dramatic experience.

In the course of 1983, the nature of dramatic experience on BBC Radio was as varied as even the most ambitious of editors could wish. At one extreme, it became the sharing, in painful but moving intimacy, of post-nuclear radiation sickness and death. Raymond Briggs' *When the Wind Blows* was an adaption of his own successful cartoon book and not therefore, strictly, an original radio play: but it became, in its new form, an experience at once as visual as the original and also, I believe, more completely involving. Jim and Hilda on the page went from humour to sadness; in the head, their fate was overwhelming.

At a bound, radio drama crossed four thousand years and became, in another adaption, an experience of Bronze Age Greece. *The King Must Die* and its sequel *The Bull from the Sea* were Mary Renault's version of the story of Theseus, a natural for the medium in many respects, not only in the scope of the story but its demand for the equivocal bull: bull or man; storeroom or labyrinth; earthquake or god? Radio revels in such ambiguous images.

Drama derived from other sources was responsible for some of the most haunting of images of this, as of any year, whether they were of England at the time of Mrs Gaskell's *Wives and Daughters* or of contemporary Moscow and the seedy hotel detective in Anthony Olcott's *Murder at the Red October*. Thrillers came back into their own in both single play and serial form during 1983. Murder was once again strictly for pleasure, and spies spied energetically and enigmatically world-wide.

Nevertheless, as in previous years, the dominant element in the drama output was new writing for the medium, and above all the single plays on Radio 4 and Radio 3. In this field alone, of some 500 broadcasts, over 300 plays were plays conceived and written especially for radio (I exclude even the most imaginative scripts which had their origin in another form, such as a short story or part of a novel). As usual, a surprisingly large number came from writers making their debut as dramatists — 62 of them were actually broadcast in the course of the year and more were bought for future transmission.

A notable revival during the year was the *Radio Times* Drama Awards, first presented seven years earlier, for the best original scripts for both radio and television, and one of the richest prizes in drama. The 1983 entries for radio added a further 612 to the 10,000 submissions considered by BBC Radio each year, and have already resulted, not only in the purchase of the joint winners, but of more than 20 of the runners-up. Amongst them, it seems likely that there will be contenders for the Giles Cooper Awards of 1984, the year in which most of them will become eligible. At the end of 1983, we had the welcome news that *Radio Times* wish to make their Awards a biennial event, so that it will not be long before substantial prizes are again on offer for writers of talent.

In Britain at least, then, there is continuing commitment to radio drama, and especially to the radio dramatist. That commitment was underlined in the summer by the opening at Maida Vale of a new drama studio, named in memory of Val Gielgud who was for more than thirty years head of the BBC's Radio Drama department. A studio is of course merely an aid to the creative process but it's worth remembering that it is a very sophisticated and expensive one, which repays its investment only over a period of years. At the opening ceremony, Richard Francis, Managing Director of BBC Radio, announced that more drama studios were planned for the BBC's proposed new radio centre, and in recent months there has been much activity to make sure that what is built will be the best.

As one who worked for Val in the department he was largely responsible for creating, and who was later privileged to be a friend, I think he would quietly have enjoyed his association with our most modern studio. But I think he would also have wished to remind us, wisely, that in radio the writer comes first; that radio is a medium for telling stories and creating pictures in the listener's head, and that whatever technical aids we develop to assist us, it is in the unique imagination of individual writers that our true wealth and our most important investment really lie.

Richard Imison
(January 1984)

THE GILES COOPER AWARDS: a note on the selection

Giles Cooper

As one of the most original and inventive radio playrights of the post-war years, Giles Cooper was the author that came most clearly to mind when the BBC and Methuen were in search of a name when first setting up their jointly sponsored radio drama awards in 1978. Particularly so, as the aim of the awards is precisely to encourage original radio writing by both new and established authors — encouragement in the form both of public acclaim and of publication of their work in book form.

Eligibility

Eligible for the awards was every original radio play first broadcast by the BBC domestic service from December 1982 to December 1983 (almost 500 plays in total). Excluded from consideration were translations, adaptations and dramatised 'features'. In order to ensure that the broad range of radio playwriting was represented, the judges aimed to select plays which offered a variety of length, subject matter and technique by authors with differing experience of writing for radio.

Selection

The producers-in-charge of the various drama 'slots' were each asked to put forward about five or six plays for the judges' consideration. This resulted in a 'short-list' of some 30 plays from which the final selection was made. The judges were entitled to nominate further plays for consideration provided they were eligible. Selection was made on the strength of the script rather than of the production, since it was felt that the awards were primarily for *writing* and that production could unduly enhance or detract from the merits of the original script.

Judges

The judges for the 1983 awards were:
> Martin Esslin, Professor of Drama, Stanford University, California and ex-head of BBC Radio Drama
> Nicholas Hern, Drama Editor, Methuen London
> Richard Imison, Script Editor, BBC Radio Drama
> Gillian Reynolds, radio critic, *The Daily Telegraph*

TIME SLIP

by Wally K. Daly

To Martin Jenkins
luminary and friend

After the successful first production of his musical *Follow the Star* at Chichester Festival Theatre, Wally K. Daly — on the advice of doyen radio producer Alfred Bradley — applied for and received a writer's bursary from the Arts Council of Great Britain, and at the age of 33 left the security of his job as Chief Electrician at the Duke of York's Theatre in London's West End, to try his hand as a full-time playwright for the six months the grant would support. April 26 1984 saw the ninth anniversary of his departure from the Duke of York's Theatre; but feeling still that he may yet at some point need to return to earning an honest living, the tools of his former trade are still retained.

At the time of going to press he was working on a new comedy series for Yorkshire Television; a new comedy series for BBC Radio; a 'Play for Today' for BBC Television; completing the final part of his Saturday Night Theatre science fiction trilogy *With a Whimper to the Grave* prior to writing two further episodes for the *Juliet Bravo* series.

He lives in West London with his wife Pauline, son Adam, 17, and 13-year-old daughter Samantha.

Time Slip was first broadcast on BBC Radio 4 on 3 May 1983. The cast was as follows:

PAUL	Paul Daneman
FAYE	Norma Ronald
FRANK	Donald Hewlett
MARGARET	Gwen Watford
HAROLD	Eric Allen

Director: Martin Jenkins

A heavy front door is heard opening, then closing.

FAYE (*distant*). Hello?

PAUL (*calling cheerily*). Me darling . . .

   His coat being removed.

   . . . just managed to catch the ten to.

FAYE (*distant*). Thought you were early.

PAUL. Where are you?

FAYE (*distant*). Kitchen.

PAUL. Sherry?

FAYE (*distant*). Please.

PAUL. Usual?

FAYE. Mm.

PAUL. I'll bring it through.

   PAUL *hums to himself passing through to the lounge: 'The runaway train' with occasional train 'hoots' interspersed with dialogue.*

   (*To himself.*) A whisky to celebrate perhaps . . . yes that's the answer. Just a little one. (*A drink being poured.*) . . . well perhaps a little bigger than that . . . (*The drink is topped up.*) . . . There we go. A toast to — just once in a lifetime managing to catch the early train instead of the late one, and not ending up drinking quick halves in that grotty little station buffet . . . (*The sound of drinking.*) . . . cheers Paul. (*In reply to himself.*) Cheers . . . Ah delicious! Must say Frank got somewhat out of breath. Two minute run left him panting like a passionate Parisian — the mole on his nose all shiny with the exertion — dreadful sight.

Right, what else? Ah sherry for Faye . . . (*The decanter is opened, a drink is being poured.*) . . . there we go. (*The lid goes back on the decanter.*) . . . top up for me. (*A drink being topped up.*) Whoops! . . . and chug off to the kitchen.

Sound effects: a door opening. PAUL, humming the train song gently to himself, chugs through the hall to the kitchen. The kitchen door opens. FAYE is hard at it.

So here you are my little lovely — slaving over — what is it?

FAYE. You're early.

PAUL. Welcoming peck for hubby perhaps? (*Sound effect: peck.*) One sherry, medium, as per request.

FAYE. Put it down a sec . . . (*The sound of the oven opening, dishes being adjusted etc.*)

PAUL. Mmmn! Smells good.

FAYE. Lamb casserole. Just about ready. I'll turn the potatoes down a touch to keep them ticking over. Then I'll be with you.

PAUL. Have a good day?

FAYE. Pretty average — right there we are. All ready when we are. Ah! Thank you. Cheers.

PAUL. Cheers.

They sip.

FAYE. You?

PAUL. Me?

FAYE. Good day?

PAUL. Oh yes — super — not only caught the early train . . .

FAYE. Wonders will never cease.

PAUL. . . . but, also got a marvellous new gadget up from our Midlands branch . . . (*Grandly.*) An 'Electronic Agitator-stroke Molecule Adjuster'!

FAYE (*laughs*). A what?!

PAUL (*normally*). Electronic Agitator-stroke Molecule Adjuster — amazing bit of equipment — pity it doesn't work.

FAYE. What is it supposed to do when it does work?

PAUL. Oh just a new bit of jiggery pokery, nothing to strain your little head with, my darling.

FAYE. Thank you very much!

PAUL. Pleasure! But if it came off it would be fantastic. Instant reproduction of three-dimensional objects — unbelievable.

FAYE. Sounds it! Your little 'Playtime Electronics' seems to be getting rather above itself these days — when you and Frank started I thought the idea was all down to computerised toys for tired 'Execs', metallic balls bouncing about.

PAUL. Sounds like a robot's dinner dance.

FAYE. And things going ping-pong — and you both having lots of free time for your golf. Now it sounds like science fiction gone mad.

PAUL. Well it's the take-over that's done that. And a good thing too. No job for a grown man, playing around with electronic pastimes all day.

FAYE (*laughs*). Come off it — you both had a whale of a time at first. What with Frank's electronic wizardry and your penchant for practical jokes, you couldn't fail to make the best adult toys in the country. Trouble is — since you've been taken over it seems to me a lot of the fun has disappeared. No sign of a practical joke for weeks — thank the Lord.

PAUL (*laughs*). You're right of course. Always the same when the big boys move in. Gobble you up and suddenly — bingo! There's hardly time for the occasional giggle, never mind golf! Mind you — they are sending us the most amazing stuff to test these days.

FAYE. Beats me how they can afford it.

PAUL. Apparently they're awash with government subsidies. In fact, wouldn't surprise me if the 'Min of Def' wasn't hidden away behind it all.

FAYE. 'Min of Def'?

PAUL. Ministry of Defence. Spies. Secrets. 007 what have you.

FAYE. You're joking! Aren't you?

PAUL. Yes I am actually — but it's a thought. The thinking is that 'New Technology is the Keynote to Prosperity' — that today's toy will be tomorrow's goldmine and certainly our own little goldmine's never been so well off. Same with Frank and Margaret I shouldn't wonder. How's your sherry?

FAYE. Fine.

PAUL. Shall we go through, have a sit down?

FAYE. If you like. Dinner is ready, mind.

PAUL. I'd like a quick shower first. Bit tacky from leaping after trains.

I'll just have a quick sit and finish this, then shoot off and shower — all right?

FAYE. All right — but no forty winks — otherwise it'll be burnt lamb in cinder sauce.

PAUL. No forty winks.

FAYE. Good. I must admit that 'your end of a hard day' snoozes have been getting a bit 'drooly' of late.

PAUL. Really?

FAYE. Yes!

PAUL. Forty winks are herewith cancelled for ever.

FAYE. Good.

PAUL. Until the next time that is.

FAYE. I might have guessed.

PAUL. Afterwards we can settle down to a bit of Brahms followed by some mindless telly. Yes?

FAYE. No. I told Margaret that if she and Frank fancied a hand of bridge later on, they could pop over after dinner.

PAUL. Cancel the Brahms and the mindless telly, Paul!

FAYE. I could ring up and tell her not to bother if you'd rather not.

FAYE *and* PAUL *sit.*

PAUL. No, no it's fine. Frank took 25p off me on the train at gin rummy, give me a chance to get my own back. Bit of aggressive bidding at 1p per thousand points should soon sort my twenty-five out.

FAYE. No penalty doubling!

PAUL. Right. Might even get the chance to talk about this new gadget at half-time coffee, now we've both had time to think about it that is.

FAYE. And no work talk!

PAUL. Right. Your word is my command. Top you up?

They both sit sipping.

FAYE (*a bit sharply*). No I'm fine.

Pause.

PAUL. I'll just have a smidgen more — then I'll be off showerwards.

PAUL *is heard pouring a drink.*

FAYE. That's at least three smidgens since you came in.

PAUL (*poetically*). But such small smidgens — wouldn't sate three pigeons. I say that's rather good don't you think?

FAYE. About average. I should imagine the Poet Laureate is safe for a while yet.

PAUL. Ah.

FAYE. What exactly is this 'Thingamybob' supposed to do?

PAUL. 'Thingamybob'?

FAYE. You know the 'Whatdoyoucallit' that they sent you from head office.

PAUL. Oh that. The (*Grandly:*) 'Electronic Agitator-stroke Molecule Adjuster' . . .

FAYE. Terrible name, sounds positively unhealthy.

PAUL. Oh . . . the idea's healthy enough. Mass-production without labour: the manager's dream and the shop steward's nightmare. For want of a better word, it's a three-dimensional copier that reproduces solid objects ad infinitum. Shove in original — push button — out pops a perfect copy. If it worked it'd make the Industrial Revolution look like a small local disturbance.

FAYE. Well, in the present state of unemployment perhaps it's just as well it doesn't.

PAUL. Apparently they have had some success with small test units. Reproducing the odd plant pot exactly and what have you. Though they did tend to be a bit unstable — exploding rather loudly after a few hours or so . . .

FAYE. Sounds more like the French Revolution than an industrial one.

PAUL. We were sent the mark-2 version to look at.

FAYE. Bigger bangs?

PAUL. Bigger all round — and more fun. Put the office cat in, thought we'd see how he'd make out.

FAYE. Paul! How could you?! Little Tiddles!

PAUL (*laughs*). It's all right, no danger to Tiddles. We made sure by testing it on ourselves first.

FAYE. What?!

PAUL. Well it obviously wasn't functioning. And we were getting a bit bored by then.

FAYE. You could have been killed!

PAUL. No. It's not that sort of dangerous. Less harmful than x-rays apparently.

FAYE. Well even that's lethal for pregnant ladies.

PAUL. But I'm not pregnant dear so it's fairly safe you see.

FAYE. But supposing it had worked?

PAUL. No chance. If we'd really thought there was any possibility of that, we'd never have risked it.

PAUL *finishes his drink.*

Right, having finished off the Scotch, on to the shower. See you in a few minutes.

PAUL *stands.*

FAYE. Not too long, I'm starving.

PAUL (*distancing*). And open a bottle of red, I feel like celebrating getting home early for a change.

FAYE. That's the weakest excuse you've ever given!

PAUL (*distant*). It'll do 'til I think of something better. See you in a minute — bit peckish myself.

We hear PAUL going upstairs. He starts singing 'Figaro' in anticipation of his shower.

FAYE (*calling*). Then hurry up. (*To herself:*) Why can't he sing something else? The joke bathroom song of all time. Well better open the red and check the veg. (PAUL *sings even louder as she crosses to the kitchen.*) Paul, not so loud — can't hear myself think!

PAUL (*distant*). Sorry dear. (*Singing stops.*)

FAYE. Thank the Lord for that. (*The sound of a bottle being taken from the rack.*) . . . Now where on earth is the bottle opener . . .? (*Sound effect: cutlery drawer opening. Strains.*) Ah — good.

Bottle being opened. FAYE starts gently humming the same tune that PAUL was singing. In the distance the front door opens.

FAYE. What on earth . . .? (*Calling:*) Hello? Who's that?

PAUL 2 (*distant*). Me darling! . . . Just missed the damn ten-to again. Frank and I ran like the blazes, still didn't manage it. Had a half in that grotty station buffet while we were waiting. Faye where are you?

FAYE (*stunned*). Kitchen.

PAUL 2 (*distant*). Sherry?

FAYE (*bemused*). No thank you, I've already had one.

PAUL 2 (*distant*). Bit unusual isn't it? Before I get back? (*Jokingly:*) Not turning into a secret squiffer are you? (*The kitchen door opening.*) So here you are my little lovely! Mmmn smells good. Welcoming peck for hubby perhaps? No welcoming peck . . . Faye — what is it? You look as though you've seen a ghost? What on earth are you looking at me like that for?

FAYE (*cold, calm and furious*). This time, you have gone too far.

PAUL 2. Pardon?

FAYE. Much too far. All right. Practical jokes are one thing. I've put up with a lot of yours over the years, but this time you've gone *much* too far.

PAUL 2. But . . .

FAYE. I think it's despicable the way you've engineered the whole thing to make a fool of me.

PAUL 2 (*bemused*). But I haven't said a word except 'Hello' and 'Would you like a sherry?'

FAYE. And it's not even April the first. What has possessed you? Even going to the extreme of singing too loud in the bathroom to prove you were showering. It's just too much, Paul . . .

PAUL 2. Faye — what are you going on about?

FAYE. How on earth did you get down the stairs so quickly without me hearing you . . .

PAUL 2. Faye! I honestly don't know what you're talking about!

FAYE. I know — you climbed down the drainpipe into the front garden didn't you?

PAUL 2. Did I?

FAYE. So you admit it? Well let me tell you Paul Williamson, if you've laid one foot on one flower, you're in for the quickest divorce on record.

PAUL 2. Should I go out and come back in so we can start again in a friendlier fashion?

FAYE. We've already had the 'friendlier fashion' Paul. This is the 'Don't try to make a fool of me' version.

In the distance PAUL 1 *starts singing again.*

PAUL 2. I still don't know what on earth you are going on about Faye. And incidentally who is *that* singing my tune?

FAYE. Don't you realise Paul you've been twigged, rumbled — the joke's misfired. All that talk of putting little Tiddles in the reproducer after you and Frank getting in to try it first; just to set me up to make a fool of me with one of your silly practical jokes. Well this time you've gone too far. (*Pause.*) You may well look aghast Paul. I am truly, truly furious with you!

PAUL 2. How did you know about the reproducer and little Tiddles? And who the hell's singing upstairs?

FAYE. Stop it Paul; or I shall walk out of that front door and never come back! (*The distant singing stops.*) See your tape has finally run out. Now will you confess?

PAUL 1 (*distant*). Where are the clean towels darling?

FAYE (*calling*). Cleverer and cleverer. In the airing cupboard of course.

PAUL 1 (*distant*). Thank you.

PAUL 2. Now someone's calling you darling! Good Lord! You've got a man upstairs! The singing . . .

FAYE. I shall show you who that is Paul. Your bluff is about to be called. (*The kitchen door opening. Calling:*) Why not come down here little tape recorder, there's someone to see you . . . (*The door closing.*) . . . now Paul let's see you get out of this one. I may be a programmed automaton, who always calls out the same replies, so you can set up a tape recorder to answer my answers, but let's see if it can also manage to walk all the way downstairs! Now that *would* be a clever little recorder wouldn't it?

Pause. Feet are heard coming downstairs.

PAUL 1 (*distant approaching*). Faye who are you talking to? You sound a bit upset?

The kitchen door opening.

. . . and why on earth did you call me a tape recording? I was simply . . . who on earth?!

Gasp of amazement from FAYE. *Long pause.*

PAUL 1 *and* 2. Good God! It's me!

Dinner is in progress; knives and forkes clicking, wine being poured into glass — large measure.

FAYE. Thank you.

Wine is being poured into a second glass, short measure. Wine is being poured into a third glass, large measure.

No — give him the same amount you've got!

PAUL (*1 or 2*). But . . .

FAYE. No buts! I've just had about enough from both of you! I don't care which of you is real and which of you is phoney. I simply refuse to have this childishness continue — arguing over who should sit in which chair; who should pour the wine; who should cut the bread — I won't have it. Don't think I didn't notice you both eyeing the other's portions — counting the sprouts and weighing up the amount of casserole you'd been given — it has to stop *now*. You may both be involving me against my better judgement in some unwanted nightmare, but at least the nightmare is going to be a civilised one. Whichever of you two is guest at this table will be treated by whichever of you two is host at this table with the courtesy that convention requires between host and guest . . . now. Fill his glass up properly.

PAUL (1 or 2). Yes dear.

Glug of wine being poured out.
A telephone being picked up close to the microphone.

PAUL 1. Right. Having got that horrendous meal over we'd better give Frank a ring . . . (*The number is being dialled through the following*.) . . . God how awful. Two grown men fighting over every smallest detail. Ridiculous. Judgement of Solomon on Faye's part; deciding to sit at my place herself like that. (*Ringing at the end of the line as the number connects*.) . . . Mind you, I still think he got more casserole than I did. It's not right. Certainly got a couple more sprouts. I counted. Just not fair. I don't know what's happened but there's no altering the fact that I'm the real one and should get the bigger portions.

The phone being picked up at the other end.

MARGARET. Hello? Seven two four three here.

PAUL. Margaret. It's Paul.

MARGARET (*gently tight*). Hello Paul. How are you dear?

PAUL. Fine, thank you. And you?

MARGARET. Yes dear and it is me.

PAUL. Margaret? . . . Are you all right? You sound a bit . . . sloshed.

MARGARET. No, no — I'm fine thank you — I just sound like this because, I'm a bit sloshed, that's all.

PAUL. Oh. That explains it.

MARGARET. Thought it might.

PAUL. Look Margaret; sorry to bother you but . . . er . . . do you think I could possibly speak to Frank. Something absolutely horrendous has happened.

MARGARET. Oh dear! How horrendous. I'll get him for you.
(*Confidentially:*) Oh! By the way which one do you want to talk to,
the late one or the early one?

PAUL. What?!

MARGARET (*giggling*). You see — I've got two of them here —
glowering at each other, and arguing over the one pair of slippers.

PAUL. Oh no!

MARGARET. Tell you what — I'll bring whichever one has managed to
get the slippers. That's bound to be the real Frank isn't it? Ever so
possessive about his slippers is Frank. Won't even let the dog have
them. Yes hang on dear — shan't be a min. Oh! By the way, better
tell Faye to forget the bridge tonight, unless she knows how to play
it five-handed that is.

The phone being put down, a long way off.

PAUL. Faye's right; it is a nightmare! What on earth has happened?
What am I going to do? Perhaps I'll wake up soon. Yes that's the
answer. I'll wake up.

The phone being picked up.

FRANK 2. Hello — Frank here.

PAUL. Frank. Thank God it's you. It is you, isn't it? Yes of course it is.
No disguising that voice. Look tell me I'm not going crazy. You and
I did just manage to catch the early train tonight didn't we?

FRANK 2. Ah — wrong Frank, old son — it's the other one you want.
The 'unreal' one. Paul and I just missed the early train tonight, had
two quick halves in the station buffet, and caught the train ten
minutes later. Yes. You're obviously the Paul who isn't real, so I'll
just pop off and get the Frank who isn't real to come and talk to
you. All right?

PAUL. But I am real!

FRANK. Hang on old man. Shan't be a sec. All absolutely fascinating,
isn't it?

PAUL. I don't believe this. I just don't believe it. (*Praying:*) Oh Lord —
please let me wake up soon sweating in my bed saying, 'What a
terrible dream that was'. Please Lord. If you do I promise I'll be ever
so good for the rest of my life.

FRANK 1. Hello. This is the real Frank speaking. (*Pause.*) Hello? Paul?
Are you there Paul?

Long pause.

PAUL. (*quietly stunned*). Yes — I'm here. You're not going to believe
this Frank. But *he* sounds exactly like *you*.

FRANK. I know. Not pleasant is it? And what's more he looks like me which is even worse Paul — I had no conception of how ugly I was.

PAUL. Oh I wouldn't say that . . . well — not ugly — not exactly . . .

FRANK. But why on earth did you never tell me about that mole on the end of my nose?

PAUL (*vaguely*). Mole — oh the mole. I thought you knew.

FRANK. No. I had no idea. I stopped noticing my face in the mirror years ago. Damn thing shines.

PAUL. Well you should have asked me — then I'd have told you. Anyway — what about my bald patch and my paunch? You never told me about those.

FRANK. Oh it's not that bad Paul, not really.

PAUL. Not that bad? You can see my head shining through the strands all over the place! From the back it looks as though the moon's coming out!

FRANK. I wasn't talking about your patch Paul, I was talking about your paunch. Your patch *is* bad I give you that, but your paunch is all right — considering. Hardly a patch on *my* paunch. Have you noticed the way my shirt buttons strain over my stomach? Quite repulsive. I used to think Margaret was exaggerating. I really must go on a diet soon . . .

PAUL. But what are we going to do Frank?

FRANK. Well I suppose we could make a start with a bit of jogging . . .

PAUL. No! The other thing. The fact that there appears to be two of each of us!

FRANK. Oh that.

PAUL. Yes! That!

FRANK. All absolutely fascinating isn't it?

PAUL. No it is not fascinating — it's perfectly awful.

FRANK. Ah. Better meet up perhaps — have a chat? What do you say?

PAUL. Good thinking — usual place?

FRANK. See you there in five minutes.

   The phone is put down.

PAUL. Better tell Faye I'm off.

   PAUL *crosses the hall to the lounge. A door opening. Music soft in the background.*

(*Whispers:*) Where is he?

FAYE. Who?

PAUL. Him. The other one, the 'unreal' one.

FAYE. Oh him. Well he was going to phone Frank but you were on the telephone . . .

PAUL. *I* was just chatting to Frank.

FAYE. . . . So he decided to go for a quick shower and shave, to help recover from the shock I should imagine.

PAUL. Well I hope the blighter doesn't use my razor — that'd be going too far even for a *doppelganger*.

FAYE. Don't forget dear — he does consider it to be his razor rather than yours.

PAUL. But surely you're not fooled, Faye. You can see I'm your true husband — can't you?

PAUL 2 *starts singing 'Figaro' in the bathroom.*

FAYE. Paul, you're identical in every way — as alike as two peas in a pod. When I'm alone with either of you I know you're my husband. But when you're together it's nothing less than a severe case of double vision. Quite frankly Paul, I know any minute now, I'm going to get one of my headaches and that it'll probably last for hours . . .

PAUL. Ah — good.

FAYE. Good?

PAUL. Don't you see — one of your headaches is just the answer?

FAYE. I beg your pardon?

PAUL. Sorry dear — now don't take this wrongly — but it's actually rather handy, your getting a headache at this particular moment. It would be somewhat comforting — while I'm out.

FAYE. What on earth are you talking about? Why should one of my headaches be of any possible comfort to you?

PAUL. Well — I want to go out — think things over . . . for a little while.

FAYE. You mean you want to meet Frank in the Duck and Partridge.

PAUL. Yes.

FAYE. Well why not say so?

PAUL. You all right Faye — sounding a trifle testy?

FAYE. More than a trifle Paul — teetering on the brink of hysteria would be nearer the mark.

PAUL. Don't worry — Frank'll soon sort it out — you know what a whiz he is at problem solving.

FAYE. You mean he might confirm that you're the real Paul?

PAUL. Oh that's no problem. The Frank I'm going to meet knows already that I'm the real Paul. Trouble is the Frank he's going to leave at home, just happens to think that the Paul upstairs bawling his head off is the real Paul.

FAYE. You mean . . . !

PAUL. Oh yes. Didn't I tell you? No of course I didn't. Turns out there's two Franks as well.

FAYE. Good Lord. How awful! Poor old Margaret.

PAUL. That's what I thought. I mean, for all his faults — one Frank is not bad — but two of him . . . disastrous! Excuse me a second dear.

PAUL *crosses the room and opens the door.*

(*Calling loudly.*) Will you shut up that damned noise for God's sake! I can hardly hear myself think!

Singing stops.

PAUL 2 (*distant*). Sorry old fellow.

The door closes.

PAUL. That's better. Thank goodness I don't sound as raucous as that. Makes a corncrake's croaking sound comparatively choral.

FAYE. Oh, but you sound *exactly* like that.

PAUL. Ah. Well in future feel free to give me a shout if it gets out of hand, all right.

FAYE. I do — all the time!

PAUL. I'll start taking notice from now on. Right — well I'd better be off — have my little think and drink with Frank.

FAYE. You still haven't explained why one of my headaches is such a comfort to you.

PAUL. Ah. Yes. Headaches. Well it is a little delicate. Let me see. How can I put this without you getting upset . . . I'm going out. Yes?

FAYE. Yes.

PAUL. And someone who looks not unlike me, but isn't me, is staying in. Yes?

FAYE. Yes.

PAUL. Yes, and sometimes, in more normal circumstances — in the middle of the week — like tonight for instance, occasionally, when I'm having a shower, and you're feeling in — for want of a better word — a playful mood . . . you . . . quietly creep upstairs — and you . . .

FAYE. Yes?

PAUL. Yes — through the shower curtain.

FAYE. Paul! How could you even think of such a thing at a time like this?!

PAUL. Well — I'm sorry dear — it's just that I am going out, and *he* is staying in . . . showering, and he does look like me . . . and you just might forget that he's not really me . . . so all in all — one of your headaches seems to be the most sensible solution all round.

FAYE. You mean I might — with another man! How dare you even consider such a thing?! I am a respectable married woman and have been for more years than I care to contemplate. Headache or no headache I wouldn't dream of . . . of . . . wouldn't dream of it.

PAUL. Oh good.

FAYE. And I may add; that until one of you turns up with irrefutable *bona fides* as to the genuineness of the product, headaches are from here on out a part and parcel of the daily round. To put it somewhat crudely the tap is, hereby — off!

PAUL. Ah — fair enough. Goodbye peck to see me on my way then . . .

FAYE. Certainly not! For all I know you're the imposter!

PAUL. Faye! How could you?!

The chatter of a busy pub. Distant pub door opens and closes.

FRANK. Evening Harold.

HAROLD. Evening Frank.

FRANK. Paul arrived yet?

HAROLD. Can't say I've seen him and no reaction of outraged customers finding false eyeballs in their glasses, so chances are he hasn't.

FRANK. Better make it my round then.

HAROLD. Usual?

FRANK. Please. On second thoughts make it a large one.

HAROLD. But you always have a large one.

FRANK. Well — make it a larger than large one. (*Glasses clink.*) Been a long day, and I have a feeling it's going to be an even longer evening.

HAROLD. Very good . . . there we go. Better help yourself to the ice when you've sipped it down from the brim a bit.

FRANK. Good thinking.

FRANK *sipping.*

HAROLD. Hello; here's Paul now.

PAUL (*approaching*): Evening Harold — Frank.

HAROLD.⎱ Evening Paul.
FRANK. ⎰

HAROLD. Usual?

PAUL. Please. Second thoughts, better make it a large one.

HAROLD. But you . . . Right. Have to put your own ice in mind.

PAUL. Fair enough. Oh — and could I have a packet of crisps while you're at it?

HAROLD. Crisps?

PAUL. Please.

FRANK. I'll have one as well.

HAROLD. Right. (*Pause.*) There you go. Another larger than large for Paul, plus two packets of crisps.

The sound of crisps being put on the counter.

FRANK. Take it out of that, Harold, and get yourself one as well.

HAROLD. Many thanks. Don't mind if I do.

The sound of the till opening and closing.

PAUL. Shall we go over to the other side — bit quieter?

FRANK. Yes let's.

PAUL *and* FRANK *cross the pub to a quieter corner. Both sit heavily with a sigh.*

PAUL. Cheers.

FRANK. Cheers.

Long pause. The sound of a crisp packet being opened.

PAUL. Crisp?

FRANK. I'll open mine as well thanks — starving.

PAUL. Me too

The sound of a second crisp packet being opened and a crisp being crunched.

Had your dinner yet?

FRANK. Yes. You?

PAUL. Casserole. Made for two and shared by three — starving.

FRANK. Same with me. Two small steaks cut three ways and I know I lost out on the deal. His seemed about the same size as mine but I got the fatty bit. Mind you I think I won on the sautéed potatoes. By two or three that is. Nothing too obvious. Not enough to be construed as some sort of secret message from Margaret to me or anything like that. How about you?

PAUL. Same sort of thing really. Except I think I lost out on the casserole and potatoes and possibly the sprouts. He had two or three extra . . . sprouts that is. But I did have a couple of largish ones — so it was hard to tell, on balance, who had the most. Definitely lost out on the casserole mind.

FRANK. Accidental? Or did Faye seem to favour him?

PAUL. Good gracious. I hadn't thought of that. He did keep looking at my plate and smirking.

FRANK. How awful.

Pause.

PAUL. Do I do that a lot Frank?

FRANK. What?

PAUL. Smirk.

FRANK. Only when you win.

PAUL. Oh. (*Pause as crisps are crunched.*) Doesn't alter the fact of course that food for two is not sufficient for three.

FRANK. True.

PAUL. The worst part was us both being so sickeningly nice to Faye.

FRANK. We were awfully nice as well — me and my all too solid shadow — didn't do any good. Margaret was too sloshed to notice. By the time I left, she was indisputedly seeing four of us.

PAUL. Really?

FRANK. Oh yes — eyeball reaction is always the give away — slight vacillation sets in — fascinating.

PAUL. I'll bear that in mind.

Pause.

And what do you think's happened to us?

FRANK. It's got to be the cabinets of the Electronic Agitator-stroke Molecule Adjuster, only possibility. Not sure whether it would be on the input side or the output, but the calculations I've started should finally show us.

PAUL. But it didn't work. You know that as well as I do. We got nothing out of the reproduction half of it at all. Neither did the other Paul. I've checked. So how can you possibly say it's the cabinets?

FRANK. Simple — I've already worked out that there can't have been more than a thirty-second time gap between the four of us. We just caught the train; they just missed it.

PAUL. So?

FRANK. Time slip.

PAUL. What?

FRANK. Time slip. Only answer. After we'd put ourselves through the cabinets and thought nothing had happened, it did — after we left — they were produced; exact copies in every detail, with all our memories, everything, but not able to see *us* because we were in a fractionally different time zone, and therefore invisible to them, and vice versa . . . I must say, I for one find it absolutely fascinating. . .

PAUL. It's not fascinating — it's horrendous — a gross intrusion of privacy, and pure science fiction to boot . . .

FRANK. No — this is not science fiction Paul — it's science fact. We have got ourselves involved in something so large that it makes the splitting of the atom look a mere bagatelle — absolutely fascinating.

PAUL. Will you stop keeping saying that Frank! Having my whole way of life disrupted by a live-in *doppelganger* is something I do not find in the least 'fascinating'.

FRANK. Sorry old son; you know what I'm like when it comes to this sort of situation. Always do find it . . . quite — interesting.

Long pause as they sip.

PAUL. There is of course one flaw in your theory about the 'time slip' stuff and what-have-you . . .

FRANK. Oh, and what's that?

PAUL. . . . . We *can* see them now.

FRANK. Yes — I've been pondering that one as well — I think the answer probably lies in the train journey home . . .

PAUL. You could be right there. Some damn funny things happen on British Rail these days.

FRANK. Somehow, the reality of the journey home has put us all in the same time zone. . . .

PAUL. . . . And left our poor bemused wives with a distressing surfeit of husbands.

The lounge at PAUL's *home. Music soft in the background. Chocolate box crinkles near the microphone.*

FAYE (*to herself*). A coffee cream this time perhaps. No, a pepperminty one. There we are.

The sound of door chimes, then the distant sound of the front door opening.

MARGARET (*calling, distant*). Only me!

FAYE. In the lounge — come on through.

Chocolate rustlings. The lounge door opens.

MARGARET. So there you are!

FAYE. I'm just having a pigging session.

MARGARET. I must say you look ever so cosy, stretched out there on the settee with a box of chocolates on your lap.

FAYE. I'm comforting my headache. Like one?

MARGARET. Mm. They look delicious.

FAYE. Left over from Christmas.

MARGARET. They're not liqueurs are they?

FAYE. No.

MARGARET. Oh good. In times of stress I always find myself escaping to the sherry. I'm only just recovering. Took three black coffees before I even dared contemplate swaying across the lawns.

FAYE. Bit too much escaping, eh Marge?

MARGARET. Oh yes. With a vengeance. I must say at my time of life I'm not used to these kinds of goings-on.

FAYE. Well I must say at my time of life I'm not exactly blasé about it. Honestly, they were like two kids at dinner. Arguing over every little morsel.

MARGARET. Same with mine. After dinner they had an argument about who should have the one pair of slippers. Ended up with one slipper each which must have been awfully uncomfortable. Hop-hop-hop! How they're going to sort out the pyjama situation later on is beyond me.

FAYE. I've already sorted it, thank you very much. Two beds made up in the spare room and an ongoing headache from here on out. And nobody gets through my door till this business is well and truly behind us.

MARGARET. I couldn't do that. I'd have to take a hot water bottle to bed with me and I'd be forever losing it. No. I was thinking more in terms of, one either side of me you know, but like book ends. (*Giggles.*) Could be quite fun.

FAYE. Margaret! What would the neighbours say?!

MARGARET (*giggles*). Phooey to the neighbours, that's what I say. You know — being quite honest, I think in the end I could come to terms with all this. I mean, it could actually grow on one, the thought of having two men about the house, doing the odd job and things — competing with each other to be the favourite. (*To herself.*) Except of course that Saturday nights could prove to be a bit of a problem.

FAYE. Saturday nights?

MARGARET. Just ignore me dear. I was thinking out loud for a second. Where are your boys by the way?

FAYE. Boys?

MARGARET. Pauls — one and two. (*She giggles.*) Sounds like a Vatican history lesson doesn't it?

FAYE. They're both out. One went — to have a think — after phoning Frank — and the other also went to have a think — after phoning Frank.

MARGARET. And both my Franks also went out — also to have a think after their various phone calls, from the various Pauls.

FAYE. So it's the Duck and Partridge yet again.

MARGARET. Probably. Do them good to have a little natter. Must be very worrying for them as well. I know I wouldn't like to see myself looking quite as real as that. Very disquieting.

FAYE. Well as long as they don't all end up standing at the same bar — it'd kill Harold's business stone dead.

MARGARET. Never thought about that. Be worse than those pink elephants that occasionally come out of his ceiling wouldn't it?

FAYE. Do they?

MARGARET. Oh yes. I keep meaning to mention it but I never seem to get round to it.

FAYE. Probably just as well.

Fade into gentle pub noises. Then close up PAUL *and* FRANK.

PAUL. But why wouldn't they believe us?

FRANK. With your reputation as a practical joker! The chairman would just chuckle and say, 'Ee that's a good one Paul lad — keep cheery that's the answer; and you go on checking that new "thingy" we sent you', and down would go the phone.

PAUL. You could be right. But we must do something.

FRANK. There is another option of course.

PAUL. What's that?

FRANK. Come to terms with it.

PAUL. How do you mean?

FRANK. Look, we love our work and our golf in about equal proportions right?

PAUL (*hesitant*). Well yes — but work first with me . . .

FRANK. And with me, but golf is a very close second — yes?

PAUL. . . . Well — yes.

FRANK. Right. Now because 'Playtime Electronics' is our baby, and important to us, we don't want to see it fall behind the other companies in the group — in fact we want to keep it the electronic front runner — right?

PAUL. Right — also it is a lot of fun of course.

FRANK. Of course. But so is golf. And lately we haven't been taking off as much time as we should. Everything that's come in has been too stimulating to put down, but now we've got the perfect opportunity to knock our handicap down to single figures.

PAUL. How?

FRANK. Well. Supposing there were two each of us?

PAUL. But there are two each of us.

FRANK. But permanently. Forever . . . one could be working while one was golfing, and vice versa. Half the week off — golf galore and no sense of guilt whatsoever. 'Cos the other chap's just as brainy as we are and will tackle any problem in the self-same fashion as we would.

PAUL. Sorry Frank. It's just not on, though I admit it is a very tempting thought. I cannot condemn Faye to a life of ongoing headaches, just to get myself a few extra rounds of golf.

FRANK. You could soon sort that out.

PAUL. How come?

FRANK. Turns each. . .

PAUL. What?!

FRANK. She wouldn't know the difference.

PAUL. Of course she'd know the difference! Frank! This is appalling. You are talking about *my wife*. The lady I love. The woman I have chosen to spend my life with — alone — without some awful 'undead' creature sharing her with me.

FRANK. Well just think about it, eh?

PAUL. No!

FRANK (*sadly*). You're throwing away an awful lot of golf.

PAUL. There's more to married life than golf.

FRANK. Gives you something to do when headaches are in fashion.

PAUL. Frank, I will not even contemplate it! Every time I imagine, see, that balding, paunchy, shambling wreck of a man, that is me, I shudder. Do you realise that — I shudder. How could I have let myself go to seed like that? It's nauseous. You should have warned me. You're supposed to be my friend.

FRANK. Well, what about the mole on the end of my nose . . .?

PAUL. Your mole Frank, is neither here nor there, compared with my baldness and shambling gait . . . hang on!

FRANK. What is it?

PAUL. What did I just say?

FRANK. You were grumbling on about your 'shambling gait'.

PAUL. No before that. 'Without some awful 'undead' creature sharing her with me.' Tell me what's the opposite to 'undead'?

FRANK. Well. Un-alive; equalling — *dead*.

PAUL. Exactly! Frank. I think I have just found the solution to my problem. I think, I may be intending to commit the perfect murder.

FRANK. What?!

PAUL (*whispering excitedly*). Don't you see?! I'd be living testimony that I didn't kill him. Can't you see — how could they condemn me to a life of incarceration, if they couldn't prove I was dead? Even though they had my body on a slab they'd still have me standing beside it — alive. It's fool-proof, absolutely fool-proof.

FRANK. Paul, could you really contemplate seeing yourself dead — after you'd actually done the deed yourself that is?

Long pause.

PAUL. You're right. I couldn't.

FRANK. Well, let's hope he can't either. When he gets around to considering the same idea.

PAUL. You think he may?

FRANK. Absolutely guaranteed — a mathematical certainty. *He* is *you* Paul. To the nth degree. *He* is you.

The lounge. Quiet music in the background.

MARGARET. Well, coming to terms with it seems a very valid answer to me . . .

FAYE. No it's simply not on. I don't *want* two husbands, Margaret, I want *one* husband, and I can certainly quite happily do without double washing, double ironing, double cooking, double . . . all sorts of things. It's just not on. I'm monogamous by nature; and if I am to have an unwanted extra husband thrust on me willy-nilly . . .

MARGARET. What a delightful name for a spare husband.

FAYE. . . . . I certainly don't want to end up with two who look exactly alike. Paul for all his good points does have an interesting cross section of minor irritations — from drooling while having forty winks in front of the TV — to cutting his toenails in the bathroom and not getting rid of the parings. Now, in toto they don't add up to a great deal, and are at least livable with — but instantly doubling them by having two Pauls is too dreadful to contemplate; besides which, it does of course make a laughing stock of the whole business of marrying: 'Do you take these twins to be your lawful wedded husbands?' No I do not thank you very much.

MARGARET. It's all very well going on about it in this fashion, Faye, but you seem to have forgotten that it's a *fait accompli* — we've actually *got* two of them.

FAYE. Well fate will have to be unaccomplied — one will have to go. They must quickly and amicably decide which version is real and which version is not — and the loser must go; disappear — immediately.

Fade up the interior of the public house.

FRANK. Lord knows what we'll do. Mind you we could have another double-double while we're thinking it over.

PAUL. Oh. Yes of course . . . awfully sorry — didn't see you empty. Crisps?

FRANK. Please.

Distant pub door opens.

PAUL. Right, back in a . . .

FRANK. Paul!

PAUL. What?

FRANK. Over at the door!

PAUL. Good God! It's you. Get down. Don't let him see us.

FRANK. Well I'll be damned. The cheek of the man coming into my local like this.

PAUL. God knows what Harold will make of it.

Crossfade closer to the bar to pick up FRANK 2.

FRANK 2. Evening Harold.

HAROLD. Frank.

FRANK. Paul arrived yet?

HAROLD. Hey hang on . . . we've already had this bit of conversation . . .

The pub door opens.

FRANK. Ah . . . here he comes. Last as usual. Better be my round I suppose. And talking about the usual, make mine a larger than large would you Harold. Been a long day and looks as though it could be an even longer night.

PAUL 2 (*approaching*). Evening Harold — Frank.

FRANK. Evening Paul. Usual?

PAUL. Please. Make it a large one. And a packet of crisps as well if you can run to it.

FRANK. Yes. Me too. I'm starving.

Crossfade back to PAUL 1 *and* FRANK 1 *holding a conversation in whispers.*

FRANK 1. Harold's starting to get puce around the cheeks; looks as though he's about to explode.

PAUL 1. Yes — he's not got the greatest sense of humour, particularly if he feels he's the butt of the joke.

FRANK. You're right there. Remember the electronic spiders we brought in — didn't speak to us for weeks.

PAUL. Yes. I think the best thing all round is for us to crawl through the tables on all fours over to the side door there and push off before they see us. What do you say?

FRANK. I think you're probably right.

PAUL. Off you go then. You go first. I'll be right behind you.

FRANK. Fair enough.

Crossfade to the other side of the bar.

FRANK 2 (*sipping*). Lovely, just what the doctor ordered. Two larger than large plus two bags of crisps. Take it out of that would you Harold and have one yourself.

HAROLD (*stiff*). No thank you. *Sir.*

PAUL 2 (*laughs*). No thank you! What on earth's come over you, Harold? I haven't heard those words from you in years.

HAROLD. I don't know what your game is the pair of you, but I for one don't want to play . . .

PAUL.
FRANK.} Pardon?

HAROLD. So kindly take your drinks and your crisps away from my bar . . .

PAUL. Come on Harold old son. It's your friends Frank and Paul you're talking to, what's the matter?

FRANK. I know what it is! The other two have been in here already.

PAUL. But of course! Why didn't I think of that possibility. Where are they? . . . Can't see them.

FRANK. Must have gone. Perhaps they saw us come in and slipped out of the side door.

HAROLD. I asked you gentlemen *to move.*

PAUL (*carefully*). Harold, it is probably very difficult for you to credit this, and I do admit it takes a bit of swallowing, but there are now two each of us. Two of Frank and two of me that is. And you've obviously served the other two already, which is why you're getting so miffed about us greeting you as if we hadn't seen you for days, which of course *we* haven't. And probably even thinking we might be pulling your leg or something . . .

FRANK. I can vouch for every word he's saying, Harold. There's now two each of us. Two Pauls. Two Franks!

HAROLD. Right! That's it! I've had enough. You've gone too far. Kindly give me my drinks back. Thank you very much gentlemen . . .

The sound of drinks being put on the counter.

. . . and the crisps if you don't mind.

The sound of crisp packets being taken.

FRANK. Hey! Steady on.

HAROLD. And there's your money back too. Now would you both kindly leave my establishment.

PAUL. Oh come on Harold that's a bit much.

HAROLD. Out! You're no longer welcome here. I warned you! I warned both of you. Last time when you brought those two hundred silly little 'electronic spiders' and cleared my pub in fifteen seconds flat. I said that if it ever happened again — even the slightest sniff of one of your stupid practical jokes and you were out . . . for good . . .

FRANK. Honestly Harold. Scouts honour. This is not a practical joke.

HAROLD. Out! *Out!*

FRANK *and* PAUL *go out protesting.*

Footsteps along a street as FRANK *and* PAUL *slowly stroll home. Countryside at night noises in the background.*

PAUL. Do you think *they* saw us creep out like that?

FRANK. Shouldn't think so.

PAUL. Did you see old Nellie pat me on the head as I crawled past her?

FRANK. Yes — gave me a crisp.

PAUL. Hope *they* didn't see us. Bit shifty-looking.

FRANK. If the positions were reversed, they'd have done exactly the same.

PAUL. You think so?

FRANK. I know so. At some point you've got finally to face up to the awful fact — that they are *us* Paul. They think like us; act like us; and in any given situation, they'll do exactly as we would do, because, they are . . . us. (*Long pause.*) And worse than that they believe beyond any shadow of doubt, same as we do believe about ourselves, that they are the real Paul and Frank.

Pause.

PAUL. And are they?

FRANK. It's just . . . 'heads or tails' . . . a 50/50 chance. So God only knows Paul . . . God only knows. There's absolutely no way of verifying our reality or otherwise.

They walk on.

PAUL. Funny how we spend most of our lives just accepting what we've got without even thinking about it . . .

Pause.

FRANK. How do you mean, old son?

Pause.

PAUL. Take you and me. How long have we known each other?

FRANK. About thirty years.

PAUL. All that time, and I don't even know you.

FRANK. Oh come on. You know me very well.

PAUL. I know some things you enjoy doing, yes. And I know some things that make you sad. And I know you've got a very good brain, even though you do manage to convince the world — when you're away from electronics — that you're a bit of a bumbler . . .

FRANK. I say Paul! Steady on — that's a bit strong.

PAUL. I'm not trying to put you down, Frank. I mean you're my best friend, my only friend, apart from Faye that is, and I wouldn't hurt you for the world. But I'm just trying to explain what I mean. After thirty years I don't know you, Frank. Only the surface you. The bits you let me share. The bits you show the world, that's all I know of you, the externals. But that — that other Frank, he knows you totally. He knows every little silliness you've ever perpetrated in your whole life . . .

FRANK. Good lord! How awful. Hadn't thought of the sillinesses.

PAUL. He knows your weaknesses and strengths. He knows the secrets locked in the dark corners of your mind that could cut you to the soul and finish you. He knows your 'secret' whatever that 'secret' happens to be. And Paul, that other Paul, knows mine. (*Pause.*) And I can't bear it, Frank. I'm afraid I just can't bear it.

FRANK. As bad as that?

PAUL. Yes, as bad as that. We've *got* to find a solution. Even if it turns out that we're the ones who're not real. The ones who must die. I don't care. I just can't bear anybody knowing me that intimately.

FRANK. Well surely there's no problem about finally finding out which one of us is which — which one's real and which one isn't.

PAUL. No problem?

FRANK. No, it was so obvious I hardly felt it warranted a mention. I mean by my calculations it's going to become awfully clear in the next sixty minutes which one of us is real and which isn't — isn't it?

PAUL. How?

FRANK. You mean you really don't know?

PAUL. No — it would appear not.

FRANK. Oh. Because of what you said I thought you'd twigged the fact as well.

PAUL. Twigged what Frank?!

FRANK. That there *is* a time factor involved here. You said we only met the others two or three hours ago — so it can't be very long now.

PAUL. I *still* don't know what you're getting at —

FRANK. Well — just you think about it for a few seconds — just think about it.

Long pause.

PAUL. Good God! But of course! Why didn't I remember that?

FRANK. There you are. Told you you'd get there finally.

Fade in the lounge. Music in the background. Chocolates being eaten.

FAYE. Quite amazing, we've got four men between the two of us and we can't even manage to produce a pair of them to form a four for bridge.

MARGARET. Well, a selection of them should be back soon. I can't see them all sitting around having a jolly drinking session together, not in the present circumstances that is. When one of the Franks comes back and sees our light off and your light on, he's bound to know this is where I am and come straight over with whichever Paul he happens to be with. (*The front door banging open, distant.*) Speak of the devil.

Distant animated conversation between PAUL and FRANK .

Good lord what *is* going on out there?

FAYE. Sounds like the start of an uprising. Maybe to restore the balance of power relative to sprouts perhaps.

MARGARET. Sounds more like one too many to me. I'd better just check.

FAYE. Remind Paul I've got a headache would you.

MARGARET (*distancing*). At this rate I'll soon have one myself.

The door opening.

FRANK 1. . . . All right, I'll get on to them straight away!

The phone being picked up.

MARGARET. What on earth's going on out here? Paul, you do realise that Faye's still got a headache?

PAUL. Oh hello Margaret. Sorry about that. I forgot, we'll keep it down.

FRANK. Hello dear, with you in a moment. Just making a telephone call. . . .

MARGARET. Phone call?

FRANK. . . . With you shortly.

PAUL. It's all right Margaret, everything will soon be explained. Let's go into the lounge shall we, leave Frank in peace . . .

MARGARET. Well . . .

PAUL. . . . There's a good girl.

MARGARET (*giggles*). Good girl indeed! I'm old enough to be your sister.

The door opening as PAUL and MARGARET go through into the lounge, then closing. Vaguely, in the background, through the following, FRANK can be heard speaking animatedly on the telephone.

PAUL. Hello darling. Sorry about the kerfuffle. Bit of a crisis.

FAYE. Crisis?

MARGARET. We were just wondering whether you boys would like a hand of bridge when you suddenly erupted.

PAUL. Yes crisis — being honest. It doesn't seem the most opportune moment for bridge.

FAYE. What sort of crisis?

MARGARET. Well lets hope the other two are feeling slightly more sporty.

PAUL. A crisis of time. Frank remembered about the time factor involved. That's why he's on to head office now — trying to confirm the details.

FAYE. I don't understand.

MARGARET. Surely we could at least set the table up and get the
cards out just in case . . .?

PAUL. Yes, you do that Margaret, help keep you quiet . . .

MARGARET. Paul! That's bordering on the offensive. Quiet indeed.

PAUL. Sorry! I meant to say busy, not quiet, busy. Sorry, Margaret.

MARGARET. Oh that's all right then. Busy! No offence taken. I'll
set it up . . .

The sound of MARGARET *getting up and setting up the bridge
table.*

FAYE. Paul, will you kindly explain what is happening.

PAUL. I'll tell you in just a few minutes; as soon as Frank's off the
phone, we'll both explain. No good worrying unduly until we know
for sure ourselves.

FAYE. Now I really am worried!

PAUL. I'll go and check how he's getting on . . .

MARGARET. Where're the cards Faye?

FAYE. Top drawer . . .

MARGARET. Shall I get them or you?

FAYE. . . . You get them. I can feel my headache's going to turn
vicious any moment now.

Crossfade to FRANK *on the telephone as* PAUL *approaches.*

FRANK. All right. Fine. I'll wait but be quick please. This is a matter
of life and death.

PAUL. Well?

FRANK. Wouldn't believe me at first, but he's checking the research
reports now. We'll have exact timings any second.

PAUL. God, it's creepy. It's like waiting for the guillotine to fall.

FRANK. Hello . . . Yes — I'm still here . . . I see. Between three hours,
twenty-two minutes, and three hours, thirty-two minutes, giving an
average of three hours, twenty-seven minutes over the whole series.
Any variations noted? . . . I see — none. Is there any other
information at all? (*Pause.*) What? No, no nothing else. Thank you.
You've been most helpful.

The sound of the phone being put down.

PAUL. Well? Why the sudden exclamation?

FRANK. Apparently, in a few cases in the early experiments the reproduced subjects had a tendency to go slowly transparent about ten seconds before the end. You heard the rest. The average life of the reproduction was three hours, twenty-seven minutes.

PAUL. Which means we must be already in the danger zone!

Pause.

FRANK. Exactly.

PAUL. What do we do Frank?

FRANK. There's nothing we can do. If we're the reproductions, we have to go. So let's play a hand of bridge while we're waiting to find out shall we? Mind you, with this amount of tension I'm bound to make a hash of the bidding.

PAUL. You're all right — I've promised Faye I wouldn't make any penalty doubles.

FRANK. Thank goodness for that.

PAUL. Let's go through.

The lounge door opens.

MARGARET. So there you are.

FRANK. Hello Faye. (*Kiss.*)

FAYE. Hello Frank — you all right, you look a touch . . .

FRANK. Transparent?

FAYE. No — pale — sort of wan.

FRANK. Oh. Thank goodness for that.

MARGARET (*briskly.*) Take your seats opposite partners please, gentlemen.

FAYE. Can the cloak and dagger now be removed and the secret revealed?

Seats are moved as PAUL and FRANK sit.

PAUL. Yes. I think it's about time to share our burden.

MARGARET. Cut for deal.

The sound of playing cards being spread on the table.

FAYE. Margaret, the secret of all the frenzied activity is about to be revealed.

MARGARET. I know that. But there's surely no reason why we can't go on cutting and dealing, is there? We've been hanging around waiting for a game for positively eons.

PAUL. You're right — no reason at all.

A card being turned over.

Ten of hearts.

MARGARET. Faye?

A card being turned over.

FAYE. Seven of clubs.

MARGARET. One for me.

A card being turned over.

Two of diamonds. Dash! You must be able to do better than that Frank. Ten of hearts to beat. Come along Frank, concentrate — don't just sit there moping. Take a card.

A card being taken.

Well turn it over.

FRANK. I don't have to — I can almost guess what it is.

MARGARET. Stop fooling about and turn it over.

A card being turned over.

FRANK. There. As expected. A sad looking lady all dressed in funeral black.

MARGARET. The queen of spades. Clever boy. Your deal.

PAUL (*gently*). It's a superstition, Frank. It doesn't really mean anything.

FRANK. Of course not.

MARGARET. Shuffle in front and cut behind. Come on, your shuffle, Paul.

FAYE. Right, Paul. Out with it. What's the big secret?

Cards being shuffled.

PAUL. It's quite simple really. It was Frank who remembered first, I'd totally forgotten.

FAYE. Forgotten what?

PAUL. That all the reproductions they made in the smaller version of the machine Frank and I were sent to test, were fine, except for one small thing . . . Frank — you're better at the technical side. Like to explain?

FRANK. Though brilliant in its conception, there is a flaw in the machine, namely — after a few hours . . . all the reproductions . . . blew up.

> FRANK *starts dealing. He could count the cards out as he does so:* *one, two, three, four; one, two, three, four.*

FAYE. They did what?

FRANK. They were perfect — but unstable.

PAUL. We've checked. After an average life of three hours, twenty-seven minutes, they just exploded, leaving no trace at all. And by Frank's calculation that is exactly what's going to happen to whichever pair of us isn't real.

MARGARET. You mean we could all be killed?

FRANK. *We* certainly could but not you or Faye. It's not that sort of intensity. More of an implosion really. And I doubt whether technically it could be termed a killing — simply a readjustment of atoms.

MARGARET. But which are you?! The real ones or the falsies?

PAUL (*helplessly*). We don't know.

MARGARET. But surely you must have some idea?

FRANK. Naturally, we think that we're the real ones — but so do the other two.

FAYE. Is there any sort of indication — a warning of some sort? Before it happens?

PAUL. Not much. Apparently some of the objects went slowly transparent ten seconds or so before they exploded, some didn't. Apart from that nothing. So any time now, we could both just disappear, bang!

FRANK. Of course, earlier reproductions weren't human! So we can't be scientifically certain. Paul! I've just thought of something . . .

PAUL. What?

FRANK. We must warn the others. They may not have thought it yet.

PAUL. Good lord you're right! I'll go and ring the pub, try to catch them before they leave.

FRANK. Good thinking.

> *A chair is pushed back as PAUL quickly goes to the lounge door.*

MARGARET. But can't it wait till after the bidding dear? I've hardly had time to look at my cards.

PAUL. No. Time is rapidly running out, I'm afraid.

The door closing.

MARGARET (*exasperated*). This is turning out to be the 'funniest' game of bridge I've never not had. (*Pause.*) If you see what I mean, that is.

FAYE. Well just hang on for a little while, Margaret, it could get even funnier. 'Heard the one about the exploding bridge player?'

FRANK. Faye, it is not a matter for jocularity.

MARGARET. I agree with you. Bridge is certainly one game that can do without this sort of levity.

FRANK. I'm not talking about bridge Margaret. I'm talking about the fact that there's a fifty-fifty chance that any second now Paul and I may be whipped off into a limbo, disappearing off the face of the earth unmourned, because to all intents and purposes we'd still be here. What do you do then, Margaret — you're my wife, you tell me what you do if I'm suddenly whipped off?

Long pause.

MARGARET (*not quite sure it's the right answer*). Play three-handed till the real one turns up perhaps?

FRANK. You just don't care, do you? You just don't damn well care, and never have done. Give you enough bridge, enough medium dry sherry, and enough of my income, and you'll plod along happily with any old Frank.

MARGARET. Frank! How dare you say such a thing.

FRANK. Because it's true.

FAYE. Oh come on Frank be fair. If you set Margaret up with a loaded question like that what can you expect? 'If I explode, I'm not real — what will you feel?'

MARGARET. Faye, what a lovely little poem.

FRANK. I just expect a little bit of understanding that's all. Because — if it does happen to be me — I personally will be very, very upset.

The door opens.

FRANK. Manage to catch them?

PAUL. No, 'fraid not. At least I don't think so. Harold was acting a bit strange.

MARGARET. Paul, will you please take your seat, so we can get on.

PAUL. Right. Let's see what fate has decided to deal me.

PAUL *picks up his cards.*

MARGARET *(terse).* Are you going to bother picking your cards up or not, Frank?

FRANK. Eh? Oh, sorry, miles away.

FRANK *picks up his cards.*

PAUL *(reacting to a bad hand of cards).* Well. So that's what fate thinks of me at the moment.

FRANK *(whistles his amazement).* Good lord! I know it's bad form to say anything, but with a once in a lifetime hand like this a man could die happy. I'm going to bid a straight grand slam in no-trumps, all by myself.

FAYE.
MARGARET. } What?!
PAUL.

FRANK. All thirteen tricks to be won. And no problem at all . . . *(Long pause.)* Well why are you all looking at me like that? It's virtually a lay down apart from one finesse. *(Long pause.)* Well?

PAUL. Frank, old friend, I don't want to worry you, but we can see right through you.

FRANK. I'm not bluffing, Paul. This is quite honestly the best hand I've ever been dealt in my whole life.

PAUL. I don't mean that sort of see through you, Frank. I mean — literally, see through you. *(Pause.)* You're slowly going transparent.

FRANK. Oh. *(Pause.)* You think . . .

PAUL. 'Fraid so.

FRANK *(excited).* But this is incredible! Look, you can see right through my hand! Fantastic! *(Then the reality strikes.)* Ah well. Looks as though we've lost out on this particular game then, Paul.

PAUL. Looks that way.

FRANK. I'll miss you.

PAUL. And I'll miss you too. You've been a good friend.

MARGARET. What about me?

FRANK. And you. I didn't mean all those things I said about you a short while ago — you go on looking after the real Frank, as well as you've always done . . . Goodbye Faye . . .

An explosion. Screams from FAYE *and* MARGARET.

PAUL. It's all right! There's no danger!

FAYE. Oh God how awful. Poor Frank.

MARGARET (*through her tears*). He just disappeared, his cards still lying there on the table. Oh how horrible! He's just disappeared! Oh Frank! Where are you, Frank!

PAUL. Margaret, don't upset yourself. That wasn't Frank — not the real Frank — he's on his way home now. He'll be here soon to comfort you. All this will be just like a bad dream tomorrow. And don't forget the Frank who sat opposite a few seconds ago was only a life-like reproduction — just as I am.

Pause.

FAYE (*upset*). How long do you think you've got, Paul?

PAUL. No more than thirty seconds, I shouldn't think . . .

FAYE. Thirty seconds!

PAUL. There's not much you can do or say in thirty seconds, is there? Perhaps just thank you for the memory of the life we've lived together; even though apparently, *I* didn't really live it. And most important of all, don't forget — I love you, Faye. And I always have done from the very first day we met.

FAYE. Oh Paul . . .

PAUL. It's all right — don't upset yourself.

Long pause.

FAYE. Will there be any warning . . . like there was with Frank?

PAUL. I don't really . . .

An explosion. Screams from FAYE *and* MARGARET. *Then* FAYE *starts crying, heartbroken sobs.* MARGARET *gets sufficient control of herself to try to be of some comfort to* FAYE.

MARGARET. There there, dear. Don't take it so hard. Remember what Paul said.

FAYE (*through her tears*). But they were so real, Margaret. So awfully real.

The front door opens, distant. PAUL *and* FRANK *call out 'hellos'.*

MARGARET. Here they are now . . .

FAYE. Oh thank God!

MARGARET. Dry your eyes so they can't see how upset you've been.

The lounge door opening. PAUL *and* FRANK *enter, breathlessly.*

FRANK. We raced back . . .

PAUL. Frank suddenly thought about the early small scale experiments . . .

FRANK. We thought we'd better warn the others in case they'd forgotten what happened . . .

PAUL. There's a time factor involved. You see. Are they upstairs?

FAYE. You're too late.

PAUL. You mean . . .

MARGARET. They blew up. One after the other. They blew up and disappeared before our eyes. It was awful. Absolutely awful.

FAYE. They've gone.

Quietly at first, then getting louder, PAUL *and* FRANK *start a cry of joy. They obviously end up dancing around the room together thrilled at being proved to be the real ones, perhaps chanting: 'We are the real ones, we are the real ones!'*

FAYE. (*Cutting through their celebration*). Stop it!

PAUL *and* FRANK *stop their cavorting.*

How dare you be so pleased. Two people who were as real as you are, have just died in this room. Whatever they were, wherever they are, they deserve better than this.

Long pause.

PAUL. Sorry. You're right of course — it is unforgiveable of us — but we have been under something of a strain. You see at one point we began to think that perhaps they were the real ones, and we were the reproductions.

FRANK. We had no way of mathematically confirming it. It could so easily have been us — until they exploded, that is. But sorry all the same, you're right it's unforgiveable.

FAYE. It's all right. I lost my temper, sorry. We've all been under a strain, it was . . .

PAUL. Try to forget it Faye — the nightmare's over now.

FAYE. Yes it's over.

MARGARET. Come and sit here, Frank. Let me just have a look at you.

FRANK. All right dear . . . (*Frank sits.*) We'll go home soon shall we? It's been a long day. Perhaps an early night?

MARGARET. Yes please, no more bridge for tonight.

PAUL (*sotto voce*). I'm sorry, Faye. I didn't mean to upset you. But the relief. I just can't begin to tell you.

FAYE. That's all right. You weren't to know how upsetting it was for Margaret and I. Funny thing is — my headache's gone completely.

PAUL. That's the best news I've heard all day.

FAYE (*friendly*). Paul, you're incorrigible.

PAUL. Right. Let's just look at what cards we would have been playing with, then we'll call it a night. Cor! What a terrible hand. No wonder he decided to take the easy way out.

FAYE (*warning*). Paul . . .

PAUL. Sorry dear.

FRANK (*whistles*). But look at this! I just don't believe it. With a hand like this a man could die happy.

MARGARET. Don't say that, Frank! That's exactly what 'he' said. And look what happened to him. You're just testing fate saying something like that.

FRANK. But I ask you. A lay down grand slam in no-trumps, apart from one finesse that is.

FAYE, MARGARET *and* PAUL *all gasp in unison as they react to the fact that* FRANK *is also going slowly transparent.*

FRANK. What is it? Why are you all staring at me like that? (*Pause.*) Is anybody there?

Pause.

MARGARET. You're fading, Frank. Just like the other one. You're becoming transparent!

FAYE. Paul do something. Don't just sit there — do something!

PAUL. There's nothing to do. The whole situation is impossible. The other two have gone. We must be the real ones. This just can't be happening. It can't.

FRANK. Paul what can I do? Help me! Help me Paul! What can I do . . . Paul . . .

An explosion. MARGARET *screams. Then starts to cry, sobbing loudly.*

PAUL. Good God! What's gone wrong? What on earth's gone wrong. How long have I got? Faye how long before the other Paul . . .

FAYE. Fifteen seconds. He though it would be longer, thirty perhaps, but it was only about fifteen. He just had time to tell me — that he loved me. Then there was the explosion and . . .

PAUL. I don't understand. I don't understand.

FAYE. Neither do I, Paul, neither do I. Paul . . . (*Suddenly near to tears.*) Paul . . . I love you . . .

An explosion. FAYE screams and starts to cry.

MARGARET (*through her tears*). What are we going to do Faye? Oh what are we going to do?

Long pause as Faye slowly gets control of herself.

FAYE. We've got to call the police.

MARGARET. But what can we say? That our husbands exploded before our eyes — twice! Oh Faye! I couldn't bear the looks in their eyes — least ways not while I'm this upset.

FAYE. Of course you're right but we must do *something* Margaret.

MARGARET. But Frank. What am I going to do without him? And the awful thing is, he was right about me. I do like my bridge too much, and my sherry too much come to that . . . Oh if only he'd come back — I'd never sit down at a bridge table again . . .

FAYE (*starting to cry again*). Well what about me? I was just as bad. I never showed my poor little Paul the love and affection he needed. No, it was always mid-week headaches, and feeling a bit off-colour all the time. Oh Margaret, I miss him so much already — if only he'd come back to me . . . (*She starts sobbing gently.*)

As FAYE and MARGARET share their sorrow, in the background we are aware of slightly off-key slightly drunken singing, getting quickly nearer. In the distance the front door opens. The singing is louder for a moment, then quickly hushed. Two male voices giggle conspiratorially.

FAYE (*stops crying*). What is it?

MARGARET. Ssh! Listen.

More giggling and indecipherable chat heard distant in the hall.

PAUL (*slightly tight, calling*). Hello darling! I'm back! Is Margaret here? I've got Frank with me. He's a little bit 'seven sheets to the wind' as they say, but I'm as sober as a judge so don't throw anything when I open the door.

FRANK (*calling, quite drunk*). What'o girls! Old Franky here with Pauly boy.

PAUL. Darling! Are you here darling?

The lounge door opening. PAUL *and* FRANK *enter, both feeling boisterous but gallantly trying to contain themselves.* FAYE *and* MARGARET *react as if seeing a ghost, probably clinging together for comfort.*

FRANK. So here's our little girls. Hello . . .

PAUL. So there you both are. We were starting to think no one was at home.

FRANK. Sorry Margaret. I'm a little bit in my . . . (*He hiccups.*)

PAUL. You two all right? Look as though you've seen a ghost.

FRANK. Haven't been tiffing have you girls — over two-handed bridge perhaps. Terrible game for tiffing over.

PAUL. Sorry to be late, dear. Missed the early train, missed the later train, missed every blooming train. In fact both Frank and I have a little confession to make . . . Frank?

FRANK. We have both been out celebrating . . . too well, I'm afraid, in my case.

PAUL. We had a new machine sent from head office today. Most wonderbar machine in the whole world.

FRANK. And it worked. It actually worked. Amazing! No not amazing — unbelievably fantastic! Three dimensional walking-talking — reproductions of human beings at the press of a button.

PAUL (*giggles*). Did they arrive safely by the way?

FRANK *joins in the giggling, and then recovers quite quickly.*

FAYE (*quietly*). They arrived.

FRANK. It was Paul's idea actually . . .

PAUL. Oh come on, Frank. You mustn't let me take *all* the credit old boy.

FRANK. 'Let's make a copy of ourselves to send home to the wives,' he said . . .

PAUL. 'Yes! Yes!' replied Frank.

FRANK. . . . 'and we'll have a night out on the town' . . .

PAUL. 'Wizzo!' replied Frank.

FRANK. . . . so we did.

Long pause.

PAUL. I say is everything all right, girls? You both look a bit peaky. It doesn't matter they'll blow up and disappear soon.

FAYE. They already have.

FRANK. Oh my God.

FAYE. All four of them.

FRANK ⎫
PAUL ⎭ (*aghast*). Four?

FAYE. Yes — two Pauls and two Franks.

PAUL. But . . . I don't understand . . .

FAYE. (*Interrupting, quietly controlled*). Why *two* of each of you, Paul?

PAUL. But . . .

FAYE. I mean if you had to be stupid, thoughtless, hurtful, cruel, wicked, juvenile, infantile, puerile — why not just send *one* of yourse. ves? But two of them? That's cruel beyond measure.

PAUL. Faye I honestly don't know what you're talking about.

FRANK. Neither do I.

MARGARET. Don't you? Well let me help you. Faye's talking about the fact that you wickedly sent two each of yourselves home to us so we could have the fun of watching you explode and disappear and die before our eyes . . . twice!

FRANK. Twice! But this is fantastic!

PAUL. Faye, Margaret — I know it was a terrible thing to do; but I promise you, on my honour, we only sent one each of us back.

FRANK. That's true. Scouts honour.

FAYE. But two each of you arrived.

MARGARET. Exactly. Two.

Long pause.

FRANK (*quietly*). Oh God. This can only mean one thing.

PAUL. What's that, Frank?

Pause.

MARGARET. Well come on, Frank, don't keep us in suspense.

Pause.

FAYE. Frank?

FRANK. Time slip! Again.

PAUL. But . . .

FRANK. Listen . . .

> *They* ALL *listen. Slightly drunken singing comes quickly nearer.*
> *Then the front door opens as* PAUL *and* FRANK *arrive home yet*
> *again, calling 'hellos' from the hallway.*

ALL. Oh no!

NEVER IN MY LIFETIME

by Shirley Gee

For Don
with all my love and admiration

Shirley Gee's first play, *Stones*, was runner up in the *Radio Times* Drama Bursary Award and was broadcast in 1974. Her seven other radio plays include *Moonshine*, in which Rosemary Leach won Best Actress in the Imperial Tobacco Awards, and *Typhoid Mary*, which won a Giles Cooper Award in 1979, The Society of Authors Pye Award, and received the jury's Special Commendation in the Prix Italia. She has had the great good fortune to have all her plays directed by David Spenser.

A stage version of *Typhoid Mary* was performed by the Royal Shakespeare Company at the Pit in 1983, and was a finalist in the Susan Smith Blackburn Prize. She is currently working on a stage adaptation of *Never In My Lifetime* for the Soho Poly, and on her second television play for the BBC.

She is married to actor Donald Gee. They have two sons, Joby, 17, and Daniel, 16.

Never In My Lifetime was first broadcast on BBC Radio 3 on 23 October 1983. The cast was as follows:

TOM	*from Keighley*	Robert Glenister
CHARLIE }	*from London*	Bill Nighy
WIFE		Harriet Walter
TESSIE		Aingeal Grehan
MOTHER		Kate Binchey
MAIRE	*from Belfast*	Maggie Shevlin
OLD MAN		Anthony Newlands
YOUNG MAN		Jim Reid
CHILDREN	*English and Irish*	Wendy Murray

Director: David Spenser

TWO CHILDREN. Will you give us bread and wine?
For we are the English.
Will you give us bread and wine?
For we are the English soldiers.

TWO CHILDREN. No, we'll give you none of it,
For we are the Romans.
No, we'll give you none of it,
For we are the Roman soldiers.

FOUR CHILDREN. Then we'll tell our king of you,
For we are the English.
Then we'll tell our king of you,
For we are the English soldiers.

FOUR CHILDREN. What care we for king or you?
For we are the Romans.
What care we for king or you?
For we are the Roman soldiers.

CROWD OF CHILDREN. Are you ready for a fight?
For we are the English.
Are you ready for a fight?
For we are the English soldiers.

CROWD OF CHILDREN. Yes, we're ready for a fight.
For we are the Romans.
Yes, we're ready for a fight,
For we are the Roman soldiers.

CROWD OF CHILDREN. Bang! Shot! Fire!

Cut into fierce disco music.

TESSIE. Night. Curtains blowing. And the music.

TOM. Don't, Tessie.

TESSIE: The glitter ball going round and round —

TOM: *Tess! Don't!*

> *Two gun shots, cutting the music.*
> *Silence for a moment, then quiet drum beat.*

FOUR CHILDREN. Now we've only got one arm,
For we are the English.
Now we've only got one arm,
For we are the English soldiers.

> *Cut to silence.*

MAIRE. Got them. We've *got them*.

CHARLIE. God, what a mess. I think I . . . I think . . .

MAIRE. Out. Out of here. Run. Must run.

> *Drums sound more urgently.*

TWO CHILDREN. Now we've only got one leg,
For we are the Romans.
Now we've only got one leg,
For we are the Roman soldiers.

> *Cut to silence.*

WIFE. In a cold-blooded ambush three British soldiers were shot dead
and one seriously wounded —

MAIRE. Four less to kick in the door at midnight.

MOTHER. Lured. Lured in the night.

TOM. Mum? Tess? Mum! It's dark, Mum.

> *The drums beat.*

ONE CHILD. Now we've only got one eye,
For we are the English.
Now we've only got one eye,
For we are the Enlgish soldiers.

> *Drums beat.*

CHARLIE. My head. Mine. A hole. It's not whole, my head.

TESSIE. Open your eyes and look at me.

TOM. Never again will I — never. Never.

ALL CHILDREN SHOUTING. Ears and throat and mouth and brain,
For we are the English.
Lungs and lights and balls and guts,
For we are the Roman soldiers.

Disco music and drums flare and stop. There is silence.

MAIRE (*shouts*). This is my land. I'll stand on it. Where I like.

TESSIE. My land.

CHARLIE. The army. My life.

TOM. My life.

CHILDREN (*whispering*). Bang. Shot. Fire.

ALL ADULTS. Won't you look at me? Look at me. Please look.

A bedroom at night.

WIFE. D'you want this, then, this photograph of me at Bognor, do you
want it? Vests green: three; underpants green: three; vests white:
three; underpants white: three. What's more simple? Just say you
don't want to go.

CHARLIE. I have to.

WIFE. Charlie, the worst thing that can happen is you get imprisoned.
Better than being dead. Towels. Working denims. Handkerchiefs.

CHARLIE. What would my dad say?

WIFE. Bugger your dad. How do you know when you'll be back. They
don't always put the soldiers back in their boxes when they say
they will. Don't go, Charlie. Please. You done it twice, you've done
your bit. Please.

CHARLIE. Look, love, I have to. If we weren't there they'd be at each
other's throats.

WIFE. Let them get on with it then. Socks. Pairs and pairs of socks.
Plimsolls. Boots.

CHARLIE. Don't you understand? I've got a commitment.

WIFE. It's the system, the system sucks you in. Put two fingers up to
it. Sod the commitment. What's the point of being out there?

CHARLIE. The point is what would happen if we weren't.

WIFE. What about me? What about the baby?

CHARLIE. Here, don't crush that, I've just ironed it.

WIFE. You're never here, you're always leaving me. Hairy shirts in case

it gets hairy. Combat jackets in case of bloody combat. Sewing kit in
case you have to sew your head back on —
No! I didn't mean it. I didn't mean it, God. I was in a mood. Four
months away. A third of a year. Two hundred packets of cigarettes
worth of days, thirty-two visits to the laundrette, more when the
baby comes, no more sprouts and football, blossom on the apple
treee, how can you bear it? They had this calendar up in the office
with this photograph, a sky full of smoke, a soldier striding through
a street, his rifle aimed straight at you. His eyes were hard, and so
tired. I've seen Charlie look like that. There was circles around some
dates, red circles. I thought, what's happening on those dates? They
never bloody tell us anything. And so I really gave it to him. Wish
I hadn't been like that. The day he left me. The day my husband left
me. I felt hard towards him. They're gone so quick, cup of tea, kiss
on the cheek, slam of the door. Gone.

*Drums followed by silence in an empty street in Northern Ireland
at night.*

TOM. All quiet now, Charlie.

CHARLIE. Quiet enough to hear a neckbone crack.

TOM. Funny how it seems to come from nowhere — a mob. A couple
of kids down a passage way, a couple of women picking their way
round puddles, and suddenly — hordes of them pelting bricks and
sinks and screaming. Funny, that.

CHARLIE. Not so funny last week for old Potter. Poor bastard.

TOM. By the time we got to him his legs was meat. Dog food. Could
have fed them to a dog.

CHARLIE. Pack it in, Tom.

TOM. Funny to think inside your skin you're meat.

CHARLIE. Full of hilarity, you are.

TOM. Sorry, mate. Anything on the radio net?

CHARLIE. A lot of hiss. Bit of mucking about with bricks down by
the Post Office. You all right?

TOM. I am now. My knee went dead for a bit — that paving stone.

CHARLIE. If that's the only bit of you goes dead you're laughing.

TOM. Know what gets me about here? Not knowing. Not knowing
anything. Who's he, who's she, why have they stopped, why have
they moved on? Is he walking too quickly, is she loitering? What's
in that pram, a baby or a bomb?

CHARLIE. What gets me is I'd give my right arm for a chip butty. Slabs of bread, scalding hot chips, butter pouring down your chin, can't stand the thought of it.

TOM. Thirteen thousand, seven hundred and twenty-three seconds left 'til I can get some shut-eye, lop off six seconds for how long it took to say it, that's thirteen thousand seven hundred and nineteen, no, eighteen . . .

CHARLIE. *One* Paddy Whack, *two* Paddy Whack, *three* Paddy Whack . . . my spine's gone numb. What's the time?

TOM. Twenty-fifty precisely.

CHARLIE. She'll be getting a cup of coffee ready for the news. Hundred to one she lets the milk boil over.

TOM. Wouldn't see me getting stuck with a woman, not in this lark.

CHARLIE. Wait 'til you grow up. All quiet on the western front. Street's licking its wounds. The stars are good tonight. Night's the time, eh?

TOM. Yeah. Night's the time. At least it's juicy when you're doing something.

CHARLIE. Can make the old eyelids twitch a bit. The silence.

TOM. You scared then, Charlie?

CHARLIE. No. Down to you, isn't it?

TOM. Not always.

CHARLIE. Always. Something silly happens, it's your fault. It's like driving on a motorway, you slam into the back of someone in a pile-up, that's your fault. You got to pay attention, do your job, keep the machine in tune.

TOM. Must have been hundreds of the sods out there.

CHARLIE. Look mate, your frontal lobes are ticking over, the heart's beating nice and steady, eyes flashing everywhere, you keep your stupid head behind the wall at the appropriate moments, it's a doddle. Right, let's belt up and listen. Listen for trouble.

A mixture of drums building to a sharp knock.

WIFE. In a way you're always waiting for it. The knock on the door. Or it waits for you. When I saw the three cars pull up outside the flats I knew. You don't get cars like that for nothing. I prayed it'd be someone else, not us, Lord, not Charlie, don't knock on my door. It's a dreadful sound, slams your heart shut. But of course it was my door. Turned out it wasn't the worst after all, only the second worst — really badly wounded. They'd been to see the sergeant's wife and the corporal's wife with the worst and made the round trip of it. The colonel and the colonel's wife, the families officer and the families officer's wife, and the chaplain — his wife didn't come, perhaps he hasn't got one. It was the colonel said it — I think it's a question of rank. 'I'm very sorry to have to tell you', he said, and he was, you could tell. He said a lot about what a fine soldier Charlie was.

CHARLIE REMEMBERED. A soldier must be fit. He must be smart in his bearing.

WIFE. He said not to worry, he'd lost a lot of blood but he was holding his own. He'd got a bit of a head injury, he said.

CHARLIE REMEMBERED. A soldier must be able to give orders and have good knowledge of military weapons.

WIFE. I wondered how it had happened — you don't like to ask, to be a nuisance. I don't recollect it all, but he said Charlie was splendid, I remember that. Putting up a splendid fight.

CHARLIE REMEMBERED. A good soldier must show his enthusiasm in everything he does. He understands the military side of life.

WIFE. The colonel had a skin like grapefruit. Some stitching had come away from his glove and I thought why doesn't your wife mend that for you and I wondered if they were doing a good job mending Charlie's head. There was two wasps dead on their backs in the dust on the window sill, I wanted to give it a wipe but I thought perhaps I shouldn't, not while they were talking. I couldn't take in everything they said, but I know they didn't tell me all of it. I had to find that out from the papers. That's how I found out about the girls.

The disco music comes up behind the end of this speech and dies down behind the next.

MOTHER. Well, she was up to her eyeballs in it, wasn't she. I never knew. I never knew at all. I sit here in her room and I think and I think — where did I go wrong? Did I do a good job on her, you know, did I really do my best? They're calling her all these fancy words — martyr, murderer, heroine, slut. And all the time it's only Tessie.

TESSIE REMEMBERED. Only me.

MOTHER. I'm thinking of it night and day, it's like an abcess drawing. I think maybe I'll never in my lifetime see her in this room again and then I don't like to stay here, but what if I start to forget something about her, some detail, we fade. I've packed her things away in boxes 'til they let her out —

TESSIE REMEMBERED. Let me out.

MOTHER. — God willing I'll be spared 'til then. Some days you're scared to open your door for fear you'll be shot like a dog on your own front steps. You can feel the fear out there, slinking like the wild dogs that sniff in the gutters, clamouring like the wild children hunting in packs through the mounds of stinking rubbish. I've just kept out the three things, her silver baby spoon, the photograph of her in summer splashing in the sea, the little blue glass horse she loved. I always touch them, always, before I lie down to wait through the shaking nights for sleep. Tessie. My Tessie.

TESSIE REMEMBERED. Mother?

MOTHER. She had everything before her. I loved every hair on her head.

TESSIE REMEMBERED. Mother!

MOTHER. Every hair. As for the men that are dead, the boys, well. I'm sorry for their families, I truly am. I wouldn't wish this grief on anyone. But I suppose they were for violence, weren't they? Violence is what they got.

Beating disco music with urgent drums.

WIFE. The injured soldier was found in an alley riddled with bullets.

CHARLIE. Thought I'd bought it, but no. Just my legs, the legs, they don't move. Don't move! DON'T!

WIFE. Riddled with bullets, riddled with bullets.

TESSIE. The glitter ball going round and round. The music, the beat of it, and the circle of glitter. And then. And then. TOM!

MOTHER. In the night. Just before midnight.

MAIRE (*shouting*). I damn them all.

Ferocious drumming, — mixed orange and green rhythms.
TESSIE's room at night.

MAIRE. They used the dogs today. The first rain for a week and I have to be stuck lying under a car in the community car park.

TESSIE. You're all over oil and muck, Maire.

MAIRE. I know it. I'm squeezed in there and I can see this squaddie's boots go squelching by and hear the clank of all that fearsome metal they have on.

TESSIE. I wondered where you were. I looked for you down the disco.

Were you up to something?

MAIRE. I wish I was. Tessie, I'm frantic for something to do. All they give me is the phones and running round a bit, the rest of the time I just sit there in the H.Q. flicking my fingers. Honest to God, I can feel myself crumbling and flaking like some old bombed-out building.

TESSIE. What was all that at the car park, then?

MAIRE. Oh, that was nothing. Well, it wasn't bad. A wee bomb. I must say, it looked terrific after, a proper reeking shambles. But I wasn't really part of it. I'm bound to get some action soon; if I could just find some way to prove myself.

TESSIE. I like your hair, Maire, it suits you red.

MAIRE. It's getting to feel like a tow rope, changing the colour all the time. It's wise, though. One of my rules. Rule one, lie like a stoat, lie about everything, name, address, age, colour of hair, job, not that there is one, where I was last night. Especially where I was last night.

TESSIE. Where were you, then? At Brendan's?

MAIRE. Ask too many questions, they'll find you in an alley. Rule two, know where the doors and windows are. Watch them. Watch the shadows. It's no use switching off, you might need to remember this yourself one day.

TESSIE. God forbid.

MAIRE. Rule three. Know where your shoes are.

TESSIE. Why?

MAIRE. Jesus, you're thick. In case you need to run.

TESSIE (*in her mind*). Maire. Absolute best friend and sometimes enemy. Arms round each other, secrets, spitting races on the roof, hiding dog-ends in candlesticks, squealing in vests under the summer hosepipe. In school while I wrote my name endlessly on rulers and desk tops, Maire was learning how to hate and how to burn a bus. An uncle in the Crumlin, a brother in the Kesh. More years, another school, our desks together still. I dreamed of seven red-haired sons and dancing, lights and glitter. Maire, a brother fled across the water, another brother shot, frowned over other things entirely.

MAIRE YOUNGER. 'It is not those who can inflict the most but those who can suffer the most who will conquer.' Ta me, ta tu, ta se, ta si . . . the right mixture of potassium nitrate, sugar and battery acid can cause a startling incandescence . . .

TESSIE. She had that incandescence herself. Maire, my mother said, mark my words, that one'll either end in jail or dead in a box.

TESSIE's *room.*

TESSIE. Is your brother working?

MAIRE. Sometimes. He doesn't really know if there's anything doing 'til he gets there. If there's nothing on he's at the pub. He'd like to find something steady now and settle down — I think he'd do anything, anything at all. He's changed a lot. Hasn't two words to say for himself. Did I tell you Michael's in the Kesh?

TESSIE. Since when?

MAIRE. It'll be two months now. He'll be starting to smell, and you know what a one he was for a clean shirt. Still, it's his choice, he's there because he wants to be. I will say for my family, they keep on coming. Conn's at it now. You should see him in action, hops like a flea over the barricades.

TESSIE. He always was a nimble little bugger. Is he six now, or seven?

MAIRE. Eight. They grow when you're not looking. He belts down the street, always in the lead, taking on the whole bloody army. Only thing is, his aim is hopeless. He misses by a mile.

Cold, clear children's voices gathering in speed and savagery.

ENGLISH CHILDREN. How do you sink an Irish submarine?
Knock on the door.
How do you stop an Irishman from drowning?
Take your foot off his head.
What do you do if an Irishman throws a grenade at you?
Take out the pin and throw it back.
What's black and crispy and hangs from the ceiling?
An Irish electrician.
Why do Irishmen smell so bad?
So even the blind know where they are.
What do you call a pregnant Irishwoman?
A dope carrier.

MOTHER. It's a hard world to raise children in, right enough. It's all gone on so long, so long. When I was six my father, Tessie's granda, before he had his foot blown off him, took me one night to the moonlit square. A group of silent men and women made a path for us, and there, laid to attention on the stone, was six dead naked men, each with a neat hole in his head where the bullet had gone in. Each man's still, grey forehead had a cross carved deep, they'd crosses gouged out of the soles of their feet. 'You will remember this,' he said. I do. You try to forget, but you can't. 'These are the great unconquered dead,' he said. 'They carried the torch of freedom. Now someone else must pick it up and keep it flaring. There's plenty waiting for the chance. Dragon's teeth,' he said.

Children hum the song beneath while the questions and answers are called out amidst laughter. Children mimicking gunfire.

IRISH CHILDREN. Sten gun?
  Clumsy. Good for close range.
  MI carbine?
  Light and handy for street fighting.
  Astra 357 Magnum?
  Great for assassination.
  Smith and Wesson?
  Take it with you shopping.
  Thompson sub machine-gun?
  Borrow one off your granda.
  Armelite?

  A burst of cheering over which the children whisper.

CHILDREN. Bang! Shot! Fire!

A light rustle of drums.

CHARLIE. Morning. A house to house. I hate a house to house. Always the same. Kick open the door and what do you find — fleas and wet mattresses, kids crouched freezing in the scullery, a family shivering with cold and hate. The same mother with her teapot, the same old man shuffling forward, calling you a slummy bastard, spitting a blob of Guinness at you. No young men. They're footless in the pubs or in the Kesh or back home taking jobs from us on building sites or there's a photo with a wreath around it, 'Killed in action 1971'. The same kids. Stone eyes. Silent. Makes you want to weep. Wilderness kids. Abandoned. Derelict. That's what those words mean really, not houses, little kids. Watch them, though. Soon they'll be old enough to lie on some roof, knuckles white on the trigger, squinting down their sights at you. Now they're fingering bits of tile and pebbles in their pockets waiting for you to turn your back and leave so they can hurl them at you. And all the houses up the hill to go through. Well, it can't go on much longer, can it? Not for a lifetime. Can't be many more stones left to throw.

The drums shimmer.

TOM. Morning. Roof slate's missing from number thirteen. Two houses
  bricked up. Cats sniffing round a rubbish bag. Sarge says a tea stop —
  good old Sarge. That'll be number seven. Fifteen minutes as from
  now, nine hundred seconds till I see her. Great. Usual three kids
  down the end swinging round the usual broken lamppost. Sometimes
  I feel sorry for them, the kids, there's not much for them is there?
  Mind you, there's ten year olds'd shoot their mothers if the need
  arose. I've learnt a lot. Last week one of the cooks put washing
  powder in our porridge. And Buster Jackson. One night Buster
  Jackson knocked on a door, no reply, looked up at the window and
  had a bowl of scalding soup poured on his face. Got one blind eye,
  one side of his face red pits and craters, looks like Mars. I saw this fellow
  dead in a car park, face like putty, blank, like putty, the sodding
  everlasting rain splashing off his teeth, but I couldn't even drag him
  out. You learn. He could be booby-trapped. Because they have been
  known to ram a stick of gelignite up a dead man's arse. Funny.
  I used to like the way it rained here. Soft. In veils. Friendly, my
  granny's knickers. You have to see it how it is. Number thirty-one's
  empty, been recently abandoned by the look of it. Lovely place
  for a snipe. Still, they'll see us right at number seven.

Shiver of drums.

TESSIE. Morning. Outside two boys lugging an old door towards a
  barricade, a woman in her topcoat scrubbing again at her doorstep,
  at the stains where a lad with stab wounds sagged a week ago, net
  curtains blow in a breeze, sun sparkles on car bonnets, you wonder
  what's underneath. Upstairs the thunder of granda's stick on his
  floor, our ceiling. Every time the soldiers come, the same. Roaring
  up a storm about the Brits, eight hundred years of torment, all that.
  Eyes glittering, crashing his stick. He lost his foot years ago, blasted
  off him by one of his own grenades bouncing off a wall. He can't
  abide a tea stop.

TESSIE's *kitchen, day.*

TESSIE. Sugar's in the cupboard.

TOM. Here, I like your home.

TESSIE. This place? Are you out of your wits? Three armchairs, a
  clothes horse and a budgie. You have to put a towel over your
  head to go to the lavvy.

TOM. I like it. It's a proper home. It's nice. Your mother's nice.

TESSIE. She is. She really struggles to see good in people, in everyone,
  and that's not easy here. She welcomed them, they came most

weeks, the four of them, a brick they call it, the sergeant and the corporal, the big one, Charlie, with the fancy watch, and Tom. Four soldiers chewing gum, putting their rifles down, peeling off gloves, crowding the room. Tom's long bony hands clasped round his mug, breathing the steam in, watching me. He held things carefully, that was the first thing made me look at him, as though things mattered to him. We found a secret way to talk, and then to meet.

The park, daytime.

TOM. Hey, I like those boots.

TESSIE. Best Italian leather, first time on. The fancy shoeshop had a bomb last week, only a small one, and a lot of stock was blown into the street. In two minutes flat half the women in town was there, helping themselves, hopping around the broken glass in their stockinged feet and yelling, trying to cram their bunions into strappy sandals. It was wonderful. I got these.

TOM. They're good. You should do a better job of cleaning them, I'll teach you how one day. I'd best be off. Tessie, would they mind, you know, your mother and that — you know, because I'm . . .?

TESSIE. I don't care what anyone else thinks. I think for myself. I come and go as I please.

TOM. Great. If my sergeant knew what I was up to he'd have me scrubbing out the latrines with a toothbrush for a month.

TESSIE. Better not risk it then.

TOM. Don't be daft. I'll let you know the same way when I'm free again. OK?

TESSIE. OK.

TOM. And you be careful.

TESSIE. Take care yourself.

Sinister drums in the rhythm of a helicopter.

WIFE. Then they started on the details. About his head and his stomach and his legs. How he'd had the top of his head blown right off. How he nearly bled to death in an alley. They said they'd fly me out but there wasn't much point with him in a coma they said, better to wait 'til after the baby, what with my mum's coming up special and the hospital bed booked and all like that. I let them decide really, I felt a bit dazed and all I could think was if he dies, won't it be funny being pregnant in black, perhaps I won't look so huge, and I said won't you have a cup of tea, it won't take a minute.

Best cups, strainer, biscuits overlapping in a wheel so it looked like
an occasion. And the glass on his photo didn't splinter, and the clock
kept on ticking. We talked about making bread, getting your hands
deep in the dough and pounding, pounding, and I said yeah, I know,
and I saw cracks in Charlie's brain and blood ooze from them. We
talked about holidays, Devon with all those windy little lanes, and
the amazing red of the soil. And I thought of course it's red, of
course. Because the land is streaming with Charlie's blood. It's
coming our way, flooding rivers and beneath the streets, rising
through veins of trees, it's under this carpet now and soon we'll
drown in it and won't you have another chocolate finger just before
you go. I think they thought they ought to stay. They were so kind,
didn't make you feel their rank at all, and it's a rotten job, telling
people things. Not that they told it all. Part of it. Not all.

The disco music.

MAIRE. We wait our chance in the dark.

WIFE. The murderers and one of the girls escaped.

TESSIE. Bra, tights, pants, lipstick, mascara, scent, boots, white dress —

WIFE. The wounded soldier and the other girl are still in hospital.

MAIRE. Want drives me. Want and rage.

TOM. To feel the life in her. The life.

MOTHER. Lost as if she was dead.

The park in daytime.

TOM. Daft, isn't it? I don't know where to take you.

TESSIE. We could sit on that log for a bit.

They move to the log.

TOM. That's better.

TESSIE. Are you up to a bit of devilment or what?

TOM. It's called the buddy buddy system. Cuddle up for warmth.
Could save your life in an avalanche.

TESSIE. Honest, soldiers are the biggest liars walking. I don't know
how you sleep at night. I don't.

TOM. Warmer, though, isn't it? Listen, that disco you came to, I'm
sorry about how they were, some of them.

TESSIE. It's OK. You expect it. Soldiers.

TOM. Reckon you must hate us sometimes.

TESSIE. I'm here, Tom, aren't I? Will you always be a soldier?

TOM. I like it, I'm good at it, it's what I am. You're doing something, see, something that matters. Trying to. Not over here, not dotting about in front gardens in the lupins and the dog mess, waiting for some Paddy bastard to throw a pisspot on your head. Last week a dear old lady offered me a sandwich. Very kind. Only thing was, it was full of powdered glass. Where did it get her, where's it get any of them?

TESSIE. They think you'll give up and go back.

TOM. Like hell we will.

TESSIE. How will we shift you, then?

TOM. Not like that. As long as you keep on at us, we'll keep on at you. It'll be down to some geezer, won't it, he'll decide. Then we can go home. In the end some bloke up there'll get sick of it —

TESSIE. They'd better hurry up then.

TOM. Yeah, they had. I hate the streets. I don't belong on them. Like I feel terrified in an APC or a 'copter — shut in. We're shut in on the streets. One hand tied behind our backs, like being blindfold in a shooting range. I'm an infantry man, me. Trees, mud, fields, hills, they're home to me. You look over the top of a hill — that's what it's about, Tess. What's over the top of the next hill. You see a field laid out before you and you think, what's its secrets, where can I root myself, where can I spring from? Hedges, ditches, gullies, walls, that bush, those rocks. It's a language. You have to learn it, then it's easy. Even now, here with you, I'm scanning the park, clocking. Trogging up and down the hills, Tess, climbing a waterfall in cloggy boots, big skies, good mates, nothing to beat it.

TESSIE. You feel things, don't you? I can tell. You're too good for them, the others.

TOM. They're grand lads, honest, even the sergeant for all his spit and spew.

TESSIE. You're worth six of them.

TOM. It's hard for a woman to understand. See, it's oil and water, women and the army, that's why I'll never get involved, it isn't fair. It's like you're married to the regiment, we're like a family, see. We cut our toenails, read the *Mirror*, change our socks, play clockwork monkeys on the streets, all together night and day. That sort of welds you.

TESSIE. I suppose it's better than begging for the dole or trying out the bars.

TOM. It is to me. Forty-two seconds left. Not much of an outing is it, sitting on a log getting wet feet listening to me bang on about my job. Your turn next time. And you be careful.

TESSIE. Take care yourself.

TOM. Invincible, me. Got my rabbit's foot, haven't I? See? Can't get me.

Green and orange rhythms on the drums build to a climax. Cut to silence.

MOTHER. Is it five that's died in our street? I thought it was four. Mrs McGlinty. She was coming out the cleaners with an eiderdown when the van went up. You never know, you never ever know. A leg on a lamppost, nothing else, and feathers, feathers everwhere. The Doolan boy, he'd a bullet stuck on him, no surprise there I suppose. Who else? Ah, that wild fellow rode his bicycle with the dog on a string pulled after him, he got between a soldier's bullet and a wall, but he was off his head with drugs, it was a miracle he wasn't run over long before. Peter. Wouldn't have known it was the same boy when they let him out the Kesh. Not a glimmer of his old self. His sister led him everywhere. Came to the Christmas party, didn't say a word, stood against a door and watched. Two weeks later he was dead. That's four. Oh, no, I tell a lie. The greengrocer, the one that always overcharged, he was a head job. That was the day I got my camel coat in the sale that he was shot. Fancy forgetting. Yes. Five. I told Tessie and I told her, never heed him upstairs, most of the soldiers are good as gold. All these young men, theirs and ours, they're all too used to killing, I never realised it was five.

TESSIE's *bedroom, evening.*

MAIRE. See that clump of lads down there, mooching round the lamp-post? Just now when I was coming here one of them stopped me. He said he'd killed a Brit and would I like his autograph. Disgusting. I told him, go on home you rotten little hooligan, you make me sick.

TESSIE. What are you up to then with your arms classes and street parades and hurling bombs about?

MAIRE. I'm not like that, Tessie. Congratulating yourself like that — that attitude revolts me. It's amateur. We hate amateurs, they make mess. Kids like that give us a bad name. D'you know they've still not given me anything decent to do, I'm still a messenger girl, a dogsbody. Look how red the sun is. Do you remember the morning the Brits came in — another life ago?

TESSIE. I do. That rumble. Long before you could see them. All of us hanging out the window.

MAIRE. The sun was red like now, and the sky a clear pink. Over the hill they came, rumble, rattle, battering through with their bulldozers and their APCs and their Saladins and their pigs. On and on until the sky was full of dust. I thought they were wonderful. (*Pause.*) They had Mrs Finnegan out in the night again. Didn't find a thing. In the morning she found they'd pissed on her doorstop, killed a dog and tied it to her doorknocker.

TESSIE. Could have been the Prods.

MAIRE. She says it was the soldiers.

TESSIE. Well, she would, wouldn't she?

MAIRE. You're very Brit-minded all of a sudden.

TESSIE. Not especially.

MAIRE. You're not walking out with some wee squaddie are you? I know your mother lets them in the house.

TESSIE. Whatever makes you think a thing like that?

MAIRE. You couldn't be as mad as that.

TESSIE. No, I'm not. I never would.

MAIRE. I'm glad. Because if you were —

TESSIE. Oh, Maire, most of them are decent.

MAIRE. Do you never listen to a word I say?

TESSIE. God's truth. How you are to them is how they are to you.

MAIRE. They'd beat the tripes out of you just for something to do.

TESSIE. I've no quarrel with them.

MAIRE. You're blind and deaf to the world, then. Bloody deaf and blind.

MOTHER. Should I have known? Should I have done something? I saw the way she smiled at him, the way he looked at her. But he was a good lad, not a rowdy lad. As for Maire . . . I knew she was sad and wild, you'd only to be near her and you could feel flickers of fear like candles in chapel. But to tell the truth I'd as soon not know who's who and what's what, there's enough to do here each day keeping one foot in front of the other. Better not to know. I'd a lifetime with my father, and him rancid now with disappointment. Once he took me to this steep hill, when he could still get about. He stood at the bottom staring up and there's a tree

right at the top. 'Darling', he said to me, 'It's buried up there.'
'What is?' I said. 'My Thompson. Under the tree in a rabbit hole
between the two big roots. It wouldn't take much to dig it up,
could you get me up there?' 'Father,' I said, 'I couldn't get up
myself.' 'I wrapped it in oilskin, it'll be good as new. Coudn't you
get me up there somehow — please — couldn't you?' For half an
hour we must have stood, the air as cold as glass, tears racing down
his face. 'Such a fine wee gun.' That fierce man crying. Like it was
a child buried there. Now I feel cold as glass inside, now I feel I've
a child buried.

A child hums the playground song, alone.

TESSIE. The Park. Our park. Third meeting. Trees dripping slow onto
the snow. Black trees, black lake, white grass, white birds, white
sky, everything black and white. But it isn't, it isn't. No one there in
the white space, nothing on the grass but the thin hop marks of
birds. We sat on the children's swings, there was nowhere for us to
go. They hung skew-whiff, they creaked. It was grand. We wound
them round and round, tangled the chains as tight as we could then
spun them free again, like kids, whirling through the air, our breath
a white mist when we laughed. Our hands were cold on the iron
chains, cold as the eyes of a watchful soldier, but not his eyes, his
eyes shone at me.

The park, daytime.

TESSIE. Put me down, for pity's sake. Tom, your knuckles are
enormous.

TOM. Boxing. Makes them bigger.

TESSIE. How can it? You don't get muscles on your knuckles.

TOM. Don't know. Does, though. It's a well-known fact.

TESSIE. You're very strong. Your lot had us out, last Tuesday. Two in
the morning. A tip-off.

TOM. Why?

TESSIE. No idea in the world. Probably Mrs McStay across the road,
she thinks the butcher saves us the best cutlets. They up-ended some
drawers and flung about the bedding for good measure, then they
put everything back again, went up to Granda's room, took up just
one floorboard very gingerly, just one — you've this passion for our
floors — put it back again, while Granda sat bolt upright in his old
vest spitting hell at them without his teeth.

TOM. I can imagine.

TESSIE. They were very polite. They had a grand haul, four marbles, a teapot lid, a canvas shoe and a multitude of splinters. We've been looking for that teapot lid for years.

TOM. Here, put my scarf on. You look starved.

TESSIE. Funny to think I've an uncle in the Lancashire Fusiliers.

TOM. Yeah, there's a lot of Micks — Irish people — with them.

TESSIE. Granda won't speak to him. Wants him to come out and put his knowledge to better use here.

TOM. I've an uncle can play tunes on his head. Drums his knuckles on his headbones and a tune comes out. 'Three Blind Mice', 'God Save the Queen' . . . He does it at Christmas. Every Christmas. Gets a bit boring, actually.

TESSIE. Show me.

TOM. When Irish eyes are smiling, all the world is bright and —

TESSIE. Sounds like someone banging their head to me.

TOM. Well, he's bald, see. That helps. It's to do with the vibrations in the hollows of your skull — something like that anyway. You try.

TESSIE. Raindrops keep falling on my head — OW!

TOM. Gently. You've got your brain in there you daft egg. Got to treat that head of yours with respect . . . be gentle.

Warning drums.

TESSIE's *room at night.*

MAIRE. They got Brendan. Knocked his forehead against a brick wall twenty times, one for each year of his Paddy bastard life they said. He couldn't see for four days.

TESSIE. Oh, God, Maire, not Brendan.

MAIRE. Oh, Tessie, he's a wreck. They bent his wrists and ankles 'til they swelled up like balloons. They said he'd harmed himself — Jesus. Well, we can't trust him now. He's damaged.

TESSIE. He taught me how to make a whistle from a blade of grass.

MAIRE. He taught me how to shoot. He should have been ready for them.

TESSIE. What a fearful thing for you, Maire.

MAIRE. Fearful for him. I don't think we need be scared of him, I don't think he knew too much. However strong you are you can find yourself vomiting out a confession — things you never knew you knew. God, when I saw him, it tore me up, it really did.

TESSIE. Sit down, love, have a drink of something —

MAIRE. I don't want to stay too long, there was a man out there, he stared at me.

TESSIE. Is he still there?

MAIRE. No. Maybe it wasn't anything, maybe it was just a feeling. Brendan was careless, see, and look where it got him. I'd best go home, I suppose, see my mother. Be good to sleep in a proper bed for once, have a decent meal instead of everlasting stew. I hate it there, all she does is gaze at telly, anything rather than talk, or think at all. She doesn't understand me, how I feel.

TESSIE. I suppose she thinks the boys have done enough.

MAIRE. No one has done enough. Risked enough. Bled enough. No one. Do you never get sick of sitting on your backside doing nothing?

TESSIE. Don't start on me, Maire.

MAIRE. Can you not see it's the only way to live properly? With self-respect?

TESSIE. I don't want to get involved.

MAIRE. Every one of those bastards dead is another step forward. Can you not see that?

TESSIE. No, I can't. I can't see the good of it. It all ends the same.

MAIRE. I've such a rage in me for what's been done to us. Not just Brendan, all our people. So much rage. I'd be ashamed to sit back and do nothing.

TESSIE. I don't *want* to be with you. I'm sorry about Brendan, truly sorry, but I don't believe you're doing such a grand job blowing the city to bits. Maybe you think you are. Well, if you really think it helps the glorious cause to kill a Brit or burn a house or two, you get on with it. You don't need me.

MAIRE. Violence has got to pay in the end, I've watched it. You get everything you want with violence.

TESSIE. Do you? Do you get an end to it all? Peace?

MAIRE. 'Ireland unfree shall never be at peace.'

Rumble of drums.

TESSIE. I dream of a home by a peaceful river lined with alder trees.

A cottage with whitewashed walls.

MAIRE. Soldiers show up well against a white wall at night. Easy to pick them off.

TESSIE. Wet clover and a hawthorn hedge, my man holding a mug of tea with careful hands, children with a ball instead of stones.

MAIRE. Stones.

TESSIE. I see it in a dream.

MAIRE. I see it clear.

TESSIE. Windows open to the sky.

MAIRE. A layer of plastic film, a wire cage, steel shutters, steel deflector plates, sandbags round the door.

TESSIE. Ramblers. Honeysuckle. Two bent apple trees with a washing line between. My man striding home in cloggy boots. A goose for a watchdog.

MAIRE. Sirens. Screams. The blinding flash. The scarlet gape of panic.

TESSIE. Pink and purple bog flowers. Hollyhocks. Clouds of gorse. A rock garden.

MAIRE. Always a stone to hand. Corrugated tin. The burnt out carcasses of cars. Ashes, ashes over everything.

TESSIE. The mountains, blue and brown and purple.

MAIRE. A giant bruise.

TESSIE. The river with the sun in it.

MAIRE. A wound that never heals. Ireland bleeds on for ever. Never ending. Never in my lifetime. I see it clear.

TESSIE. I see it in a dream.

The disco. Music and drums.

CHARLIE. Tried to blood — blood — no, double back down the alley. Back. Black.

TESSIE. Blind with blood. Broken.

WIFE. They were off duty and in plain clothes when they were invited . . .

TOM. Finish my time in . . . finish my time.

MOTHER. A dance. Dear God in Heaven, it was meant to be a dance.

MAIRE. Each time one less with his teeth like broken glass and the steel clank of his limbs and his eyes with their iron glare.

TESSIE. They did it. I did it. What does it matter which . . .

The music and drums mix into a single drum beat.

WIFE. When they'd gone I watched the telly with the sound off 'til the Queen came on, and I looked at her and she looked at me and I wondered what she thought about her soldiers. Then I went to bed. I thought, oh God, I'm going to dream, but I didn't. But in the morning I made myself a boiled egg and when I took the top off the shell was full of blood. It went on like that for quite a bit. I didn't cry at all. Just suddenly out of nothing, I'd turn a tap on and blood'd come out instead of water or it'd be in my slippers or there'd be someone singing on the telly and blood'd be swimming in their mouth. Only in the day, though. In the night I never dreamed at all.

The drums — sinister.

TOM. Nights I think of how they come from nowhere and they're everywhere. Round you in a circle. A sniff in the wind and they're swarming. Nervy little brats hopping like fleas over the barricades, rock-faced harridans in winter coats . . . Christ! Nights I dream of Potter. Once we had a disco and only ten birds turned up, and Potter got into drag and wore a sequin G-string. He had fantastic legs. He was a scruffy bastard, though. Farted in his sleep like nobody's business. One night we threw his bed clean out the window, then we whipped his sheets, his underpants, his socks, all his stinking gear, chucked the lot in the boiler. There was Potter, red as a post, yelling and foaming. You had to laugh. Grabbed his trousers, cut the legs off, hacked away at those legs — well, he don't need trousers now. Don't suppose he even pongs, nurses are forever washing you. Yeah, I dream of Potter sometimes.

Dark drums.

MAIRE. In the darkness I dream of my Armelite. When I've proved myself they'll give me one, I know they will. The sweet click as the parts go home, build up, oiled and easy, heavy, greased and easy. The slip of the bolt, the shine of the barrel, all going home with a sweet click. The honey shine of the grease on the barrel. The long, hard length of the barrel, a long, dark, shining length. Sliding together, making a whole. A dark, dangerous hole, a hollow depth, a dark tunnel filled with danger, spiralling on and on. Light finger on the trigger, stroking the trigger, pressing, releasing, pressing, releasing.

The drums again.

CHARLIE. Nights I miss her. What I couldn't do to a woman. A big bold tart, a strapping lass, buttocks like water melons and tits like big brass bells. I'd make them ring all right. I should have made corp, not that slag heap Hartley. There's Baby Wilkinson again, yelling in his dreams. He's a disaster. Always trapping his fingers, crying, losing bits. Poor lad, he won't last. I'll decorate the lounge when I get back. Fix that bannister. I love her, yes, I love her. But she's not here, is she? She turned away, wouldn't say goodbye, and very slow she put her arm across her face, covered her eyes. Kept her arm there and I'm stood in the doorway, the lorry's coughing and wheezing outside, the horn's pipping. Her white arm and her ragged bitten nails, her hand curled up. She bites her nails, she smokes too much, I don't know what to do with her. Perhaps when the baby comes . . . I should have made corp, I was next in line.

Whispers of drums.

TESSIE. A weakening of light. The long sigh of the Lough. No smoke charging the sky above the mountain ring, the Black Mountain and Castlereagh, Collin and Craigantlet, Divis and Craig Hill. No bin lids and whistles. Not yet. Small sounds up and down the back entries, whispers, secrets. Upstairs Granda in his dressing gown stirs up his memories like the deadly stew of potassium and detergent he brewed on his landlady's stove. He remembers the Famine, he remembers the Fenians, he remembers the Great Unconquered Dead —

OLD MAN. Daniel O'Connoll, Patrick Pearse, James Connolly, O'Donovan Rossa, Wolfe Tone —

TESSIE. And the manner of death of each one, the cut throat, the severed head, like some remember pop songs or the dates of kings, and his griefs and grudges fester.

WHISPER. They used the dogs last night.

WHISPER. Go on home. Hurry.

WHISPER. No. Not there. Orders.

WHISPER. Are they up to something? Will there be trouble?

WHISPER. You can count on it.

TESSIE. A thin shape scuttles to a wall and from its shabby pocket filled with broken brick takes out some chalk and scrawls —

YOUNG MAN. Get Yourself A Gun. Fuck The Brits.

TESSIE — and scurries on towards another pint. My mother leans her head against the cold glass of her bedroom window, closes the

curtains against the fear of night, screws her last cigarette into a
saucerful of butts, shakes out two tablets, one for panic, one for
tears, sinks to her knees before Our Lady of Lourdes and prays to
her to keep us through the night. Boots slam up the street, a quick
gust flaps and cracks the tricolour, a couple of shots across the black
river. Silence. Then —

The streets. Night. Stones rattling on railings. A chant begins.

CROWD LEADER. What do we want?

CROWDS. TROOPS OUT!

CROWD LEADER. When do we want it?

CROWD. NOW!

The chant becomes more and more insistent. Crashes. Cries.

TOM. There's a fair bit of stuff flying around.

CHARLIE. Nearly time to get in there, jump up and down a bit.

TOM. Bloody hell. That was a manhole cover just went screeching past
my ear. This lot's not going to listen to reason.

CHARLIE. Hearts and minds is over, head bashing begins.

TOM. Reckon we'll make it, Charlie?

CHARLIE. Get your head down. Do your job.

TOM. I been thinking about Potter.

CHARLIE. Well, don't.

TOM. He's had it, hasn't he?

CHARLIE. Course he's bloody had it.

TOM. Makes you think a bit. Hope it's quick, you know.

CHARLIE. Get your sodding head down.

TOM. I been wondering, Charlie, was he the one carried the radio. It's
always the one carries the radio that gets it, that's what Calvert said.
I should have asked Calvert was he the one —

CHARLIE. He copped it because he was careless. Put his great thick
stupid head over the wall.

A barrage of stones and petrol bombs.

Here we go.

Another assault.

Roll on Sunny Cyprus.

The banging of a dustbin lid in the rhythm of the chant. This is taken up, doubled, redoubled, the clang of bin lids fills the streets.

TOM. Christ, they're everywhere. Know how to make every stone count. Every stone.

CHARLIE. Brick.

TOM. Rock.

CHARLIE. Nail bomb, petrol bomb.

TOM. Blast bomb, fire bomb, acid bomb —

More explosions, jeers and screams build to a violent climax. Cut to silence.

CHARLIE (*in the mind*). Reckon you could keep your head in a crisis?

TOM. A man got hit so hard once his brains came out of his mouth.

CHARLIE. Reckon you could handle the aggro?

TOM. A man got hit so hard it sent his guts straight through his anus.

CHARLIE. Danger nicking your skin like a razor blade.

TOM. They roll up what's left, roll it up with a spade.

CHARLIE. The old heart's going pitter pat. Adrenalin flowing. Use it. Use your fear.

TOM. Black plastic dustbin bag, half full. Death sheet.

CHARLIE. Use your military skills. Use your animal instincts.

TOM. They lug it away and that's you, gone. Nothing left but a puddle of blood three foot across. Oh, God, if it happens to me, let it be quick. Not my throat, Not my balls. Straight through the head and that's it.

CHARLIE. An element of risk. That's the job.

TOM. Got my rabbit's foot. Be OK. Won't be me.

Explosion, close.

CHARLIE. Life's sweet, my sons; life's bloody magic.

TOM. Not me. Not me. Not me. Not me. Not me.

Violent crescendo of dustbin lids.

The barracks at night.

CHARLIE. Look, forget it. If they play you up you have to thump them.

TOM. She came at me. Right at me.

CHARLIE. So don't get windy.

TOM. Wasn't anything else I could do. I couldn't leave it.

CHARLIE. You'd have been nowhere. Splatted out. A lovely picture for the telly.

TOM. See, I though she was going to —

CHARLIE. That's what your baton's for.

TOM. Christ. When I hit her, though —

CHARLIE. Forget it. She was naughty.

TOM. Her face. Christ. I didn't feel like it was me, not real. Like in a film, like someone else was doing it.

CHARLIE. Here, have a swig.

TOM. Where'd you get that?

CHARLIE. Stashed away in my underpants.

TOM. I didn't think it'd be like this.

CHARLIE. Well, it is. Nobody said it's be all butterflies and birds. We're soldiers, not pooftahs, not bleeding wind-up toys, not moving targets, we're the Queen's bloody army, right?

TOM. That's right. They stop, we stop.

CHARLIE. We could have this whole pig's arse tidied up in no time if the bloke's upstairs'd see sense, let us make a proper war of it.

TOM. That's right.

CHARLIE. Next time hit harder. Kinder. Gets it over quicker. Have another swig.

TOM. You're right, Charlie. (*Pause.*) You ever met the Queen?

CHARLIE. No.

TOM. I've never seen her except on the news.

CHARLIE. Nearest I got, I did a bit of street lining for some old Blackie nabob and her car went by.

TOM. Bet she looked great.

CHARLIE. Couldn't see. He was in the way.

TOM. I'd give a lot to meet her. Probably never will. Pity, that. I think she's great. And after all, she's who we're fighting for.

CHARLIE. What I'm fighting for's to keep awake.

TOM. Christ, she screamed loud enough to crack a cliff.

A disturbed echo of the dustbin lids, faint.
TESSIE's *room at night.*

MAIRE. Did you hear about Mrs Cromerty?

TESSIE. She was hurt last week wasn't she, in that spot of trouble?

MAIRE. Yes. She had quite a bad five minutes of it.

TESSIE. I've been meaning to go and see her.

MAIRE. Do that.

TESSIE. How is she?

MAIRE. Oh, she's fine. She's at home now.

TESSIE. That's good.

MAIRE. Sat in a dark corner with a cloth over her head like a parrot in
  a cage, bones she never knew she had broken in her face. The sight
  of her makes her youngest scream.

TESSIE. She'd never say boo to a goose. How did it happen?

MAIRE. She went out to look for Seamus. She had news he'd been
  lifted.

TESSIE. He's not even Provo-minded.

MAIRE. His name's Seamus. That's enough, isn't it? She timed it
  wrong, got swept up with the crowd. This Brit went mad with his
  baton — beat her face to pulp.

TESSIE. Sweet Jesus, Mrs Cromerty.

MAIRE. Glad you're moved. Now take a look at this. Do you recognise
  him? Do you? You should. He's been welcomed in your house often
  enough. And you've been out with him, haven't you? Haven't you?

TESSIE (*a whisper*). How do you know?

MAIRE. We know.

TESSIE. That's never Tom. I don't believe it.

MAIRE. D'you think the camera lies? Keep the picture, study it, stick it
  under your pillow. An action shot of your wee squaddie.

TESSIE. Don't call him that.

MAIRE. I'll call him what I choose. Thug. Murderer.

TESSIE. He's not like that.

MAIRE. They're all like that.

TESSIE. It's blurred. In uniform they look the same.

MAIRE. If you think he's not as barbaric as the rest of them you're living in fantasy land. Open your eyes.

TESSIE. I'll not believe it. Never.

MAIRE. Go and see Mrs Cromerty, she's a family friend, isn't she? Ask her all about it, she says it helps to talk. Go in the evening when the children are in bed, she might even take the cloth off for you, let you have a look. Then go and play your games with him. Your squaddie.

WIFE. They must have seen him as the enemy, because it is a war, whatever they say. If they'd known him as he really was, gentle, helping people, mending things, if they could have seen him polishing the silver, our wedding bits and pieces . . . they couldn't have done it. Never. Never in their lives. I know he was a soldier and people think they're rough and tough. Well, he trod on a frog once that had got into the kitchen and he cried, leant on the sink and cried. He'd never knowingly hurt anyone. I don't know what soldiers do. When they're here they tramp about the place, they moan, they make a lot of noise. They never have enough dry socks, they muck about in canoes and making bridges out of rope. But when they get out there — it's such a mess out there. You know a lot goes on, you know its awkward, but you don't know what it is.

MOTHER. You live like this with the streets so tense you can reach out and touch the hate — I never knew such hate could be. I read somewhere someone on some side or other — I don't remember which and what does it matter anyway — put a man's head in a vice and squeezed it 'til he died. It's like that with my heart, with the hearts of all of us. All these people round about, taking away lives, taking away limbs. No wonder the little lads throw stones, no matter who it's at. Who's in the right of it? Where do I stand? Oh God, I'm so tired and tired and tired of it all.

CHILDREN *hum a fierce staccato version of the song accompanied by drum beats.*

TESSIE. I don't know how he got the room, but it was ours for the afternoon. He took my clothes off for me. Slowly. Sweetly. My clothes were like a diary — a memory, an event with every one. He took my coat off first and hung it on a nail. The coat I wore the day I heard they'd lifted Brendan. They bent him over a table backwards like a crab, to try to snap his back. Tom had dear bones in his spine. Knobbly. Beads in a rosary down his back.

The room.

TOM. I've missed you, love. You wouldn't believe.

TESSIE. Sometimes in the night I've held myself, pretended, you know, kissed my hand, pretended it was you.

TOM. What have you done to me? I mean, I've never —

TESSIE. I know.

TOM. It's like — I've always . . . before . . .

TESSIE. Oil and water? Love them and leave them?

TOM. It's like I've waited all my life and now —

TESSIE. I know.

TOM. You know the lot, don't you. Come here, lie here. Your knees are trembling. Little boy's knees you've got. All these little scars. You been falling out of trees or what?

TESSIE. They're nothing. A few stones.

TOM. How did it happen?

TESSIE. Years ago. I was in the wrong street. That's all.

TESSIE. Like Mrs Cromerty. She whispered it all out to me, hunched in her dark corner, poor shuddering thing, her face ruined beneath the cloth she wore. He unzipped my pleated skirt, the one I kept for best, the one I went to Mass in with my mother, although I could scarcely bear to watch her kneel and pray for the thousandth time to God for our souls and the end of living in a wilderness and please to send us a bathroom.

The room.

TOM. There's nothing to you, I can get my hands right round your waist, near enough, thin as a goal post aren't you. Don't you eat proper?

TESSIE. Our lot have always been sticks. Granda says its the famine.

TOM. That was yonks ago.

TESSIE. It left its mark on lines and lines of families. That's what he says. I don't know. The women in our family are all scrawny, that's for sure. Do you mind? Do you wish I was more voluptuous?

TOM. You'll do for me. I told you.

TESSIE. Tom. Tom, wait. There's something I must ask you.

TOM. In a minute, love.

TESSIE. But, Tom, I —

TOM. Sh.

TESSIE. The scarf with coloured squares he slid from me. Granda once when I was small wrapped his gun in it and gave it me to hold. The dark, secret, heavy, oily feel of it, even through the scarf. Then the sweater. 'Help me,' he whispered, and I held my hands above my head, like at a checkpoint, like surrender. His hands were on me, searching, but not for weapons. Imagine, a soldier's hands on me and nothing but delight. Not him, please God, not him. The sweater caught in my hair as he pulled at it, tangled. It blocked my face, and I was blind and gagged. A hood. A hood. A cloth over a face.

The room.

TESSIE. No!

TOM. What's the matter? Did I do something wrong?

TESSIE. I thought . . . I thought I was being searched.

TOM. Bloody hell. Here, put this lot on again.

TESSIE. I'm sorry, Tom.

TOM. Go on, get dressed. Made me feel great, you have.

TESSIE. Tom. I'm sorry.

TOM. Anyone search you like that, I'd bloody kill them.

TESSIE. Would you? Would you, Tom? Would you kill someone for a thing like that?

TOM. What's that supposed to mean? It's all gone wrong, hasn't it? It's this crappy room — what sort of room is this to bring your girl to? Another scabby, flea-ridden derelict — no one knows you're here, do they?

TESSIE. No.

TOM. No, nor me neither. Having to lie to meet, scared someone'll find out, and the crappy streets and the crappy people. Everything's wrong in this Godforsaken hole, milk churns explode, Christmas parcels blow you to hell, babies have bombs jammed up their nappies. This bloody country is our prison, we're locked up in it, told when to go out, gates unlocked, told when to come in again and the gates locked behind us.

TESSIE. You really hate us, don't you?

TOM. No. Not you. Not you. I'll take you home, Tess. You can breathe at home. You're free.

They kiss.

TESSIE. Tom, there was trouble in the street next to ours last week. Were you there?

TOM. Might have been. There's been a lot of shouting and screaming going on, haven't had much sleep. Haven't seen one rotten telly programme right through since I've been here.

TESSIE. A friend of ours was hurt.

TOM. People should stay indoors.

TESSIE. She was worried about her son. She went to look for him.

TOM. Just lock their doors. Stay out of trouble. Been told often enough.

TESSIE. Apparently some soldier caved her head in with his baton.

TOM. Oh?

TESSIE. Apparently he went on hitting her. I wondered if you'd seen.

TOM. Times like that there's always bags of noise, bags of confusion. No, I didn't see. Tess, let's talk about it later, let's not bother now, eh? We've little enough time. Was she hurt bad?

TESSIE. Her face is terrible.

TOM. Must have been up to something. Must have come at him, whoever did it, must have had a weapon in her hand, a brick, a rock, whoever did it must have thought . . . Not blinded, is she?

TESSIE. No.

TOM. Good. That's good. Come on, Tess, come and lie down again. Listen, to hell with the whole steaming nightmare out there, who's doing what to who and why —

TESSIE. How can we get away from it? It's everywhere.

TOM. You're you and I'm me and we're here and let's forget it and I bloody love you.

TESSIE. Oh, Tom, there can't be anything for us.

TOM. Sh. Don't think about it. Shut your eyes. We'll sort it out later, promise. I want you. Let's make love, shall we, shall we, Tess? Come on, lass, come on . . .

The disco. Music and drums.

CHARLIE. The SLR. The SMG.

MAIRE. The Armelite.

CHARLIE. The 7.62. The SLR, a hard punching weapon, penetrates but may not go straight through you. You stand a good chance in your flak jacket, unless you get it in the eye or throat.

TESSIE. Never thought I'd have a soldier's hands on me, and feel my throat ache, but not with hate.

TOM. It's like the silver bugle's calling over rooftops, like when an armada of our helicopters come flying out from clouds across the moon, like in that riot when the milk bottles turned in the sun then smashed into a million diamond splinters.

CHARLIE. The SMG, unless it hits a hard bone, tears straight through the body, rips away the flesh, leaves a hole in you like a large dinner plate.

TOM. My hands warm over her ribs, I can feel the life in her, the heat. My face in her thigh. Breathe her in —

MAIRE. The Armelite. The bullet may pass straight through the body and out the other side. A man might run on for a hundred metres before the body suddenly drops dead. There is no mess to speak of.

TESSIE. Never thought I'd feel my heart slam, but not with fear. Ah, Jesus, what delight.

MAIRE. However, if the bullet hits a large bone, it will spring off and make a pattern all through the body. Entry in the leg, say, exit in the neck. The bullet bounces round the body like a snooker ball bouncing off the cushions. Tears a man's inside out.

CHARLIE. The 7.62. Not a nice thing to be hit with, especially in the face. Imagine pushing something into, say, an orange, putting all your pressure behind it, then pushing hard and forcing through —

TOM. The face. When the smoke cleared I saw him. I seen the films, I seen the mock-ups, the dummies, but Christ, it's nothing like it. The body was dancing about, nerves still left in it. His right leg and hip was off, and three parts of his stomach hanging out. The bullet took away the eye, the nose, the mouth, the chin. There wasn't nothing left. Buster Jackson with his boiling scars. Potter's leg like meat. I cracked my baton down for them. For me.

TESSIE. Huge eyes. Thin neck. The stubble of his soldier's hair under the palm of my hand. Corn stubble. Their boots trample the corn, they wreck the fields, but let it not be him. Please God, not him.

TOM. Her face. Her scream. I'll swear she had a weapon in her hand, my life on it. Oh, Tess, I love you. Make me forget, Tess, make me forget.

TESSIE. And we forgot it all. What a room for love. A helicopter flew low overhead and made the cardboard in the broken window clap against the frame, balls of dust blew across the lino. He'd folded all my clothes into a pile — it made me laugh, so neat and soldier-like. My boots were set together side by side, pointed at the door. Rule one, lie. Rule two, know where the doors and windows are. Rule three, know where your boots are. I'd only looked away from him a moment, but when I looked back both of us were crying.

The drums and the disco music.

MOTHER. They were invited by the girls —

WIFE. Raked with bullets from six feet away —

MAIRE. The iron stench of blood.

WIFE. Riddled with bullets.

MAIRE. Red dreams. Green dreams. In the name of my green land I damn them all.

CHARLIE. She nice, nice, no — sniper. Sniper. She shouted 'Got him'.

TOM. Finish my time in four weeks, three days, six hours, six minutes and twenty-three seconds, twenty-two seconds . . . finish my time . . . finish . . .

MOTHER. A cold-blooded ambush.

Burst of drums.
TESSIE's *room at night.*

MAIRE. Where have you been?

TESSIE. Oh, are you here? To a film and then to get some ciggies. Do you want one?

MAIRE. You've been with him again.

TESSIE. I don't know who you mean. Have you a light?

MAIRE. You lousy liar. I can smell him on you.

TESSIE. Isn't that life all over? Now I've the fags and not a light.

MAIRE. Is he a lousy liar like you are? Did he say he was protecting Mrs Cromerty?

TESSIE. Will you mind your own business for God's sake.

MAIRE. You are my business, you're my friend, I care about you. Remember Maggie Ryan?

TESSIE. I've no idea. Why are you always round here anyway? Was she the one with frizzy hair? The one that always had a cold?

MAGGIE. She was the one went out with a Brit. She didn't listen — thought she knew it all. She went out once too often.

TESSIE. Your face is a picture.

MAIRE. Tessie, you've been seen.

TESSIE. I'm tired. I've started that job at the chemist's and I'm tired.

MAIRE. You've been seen. Both of you.

TESSIE *says nothing.*

You must be mad.

TESSIE. Look, I go where I please, all right? I meet who I like.

MAIRE. Not here, you don't. Not now. It's too late now. Remember Maggie Ryan? She was warned. Twice. Officially, you know? And then one day they did it. Two of them. One had a penknife and one a knife for gutting fish.

TESSIE. Sweet Jesus, I remember.

MAIRE. They took her to an alley and they slit her. First her mouth, both sides, right out to her ears. She's a grin now forever.

TESSIE. Oh, Maire, please.

MAIRE. Then they slit her again. Between the legs.

TESSIE. Oh God in Heaven.

MAIRE. They left her dumped on the Town Hall steps with a note pinned on her. Soldier lover. Oh Tessie, you're a fool. This note was given me to give to you.

TESSIE. For me? Who by?

MAIRE. They'd never give their names. I warned you, didn't I? Tessie, I didn't tell them. I swear to God —

TESSIE. What does it say, then? Have you read it?

MAIRE. I don't need to. I know. It wasn't me, I'd never have done that.

TESSIE (*opening the note. A whisper*). Soldier lover.

Warning drums.

WIFE. I got all the baby clothes and I burned them. We'd chosen them together, we'd planned it all together, I'd wanted him to be at the birth. He'd said he'd be there, and now he wouldn't be. I jumped off the table in the kitchen over and over again, I didn't want the baby, not at all. My husband went away and left me. Gone for good. If I couldn't trust him to stay in one piece why should I trust anyone? Anything? Why bring a baby into it? What makes girls like that tick, whores like that? One time there was a bunch of wives attacking the Irish wives in the NAAFI and I tried to stop them. Now I'd have started it. I didn't want to sit at home fart-arsing around, I wanted to get a gun and go over myself and do something, I wanted to kill, there's not enough of them shot. I couldn't get those girls out of my mind — what did they look like, what did they feel, did they have families, why, why, why?

MOTHER. I have a fear now when I go by myself. I'd like to be my sister. She laughs a lot, she's asked for me, I could bring the bird, her hats are lovely, you can see the sea from the top of the hill. I'd feel keen about it if I knew how. But it's a strange thing — this is my home, I've all my memories here. How I'd stand in the moonlight and watch her sleeping, wondering what would she be. Her little cold feet against my shins in the early morning, the shine and crackle of her hair when I brushed it, the clatter of roller skates. She was a demon on those skates. That last evening, what she said to me, what I said to her, why I let her go. I couldn't have known, could I? I couldn't have known. Oh God, someone'll have to help us soon.

TESSIE's *room at night.*

MAIRE. It's just a wee job. Are you not Irish after all? Are you with us? Prove it, then.

TESSIE. What do they want us to do?

MAIRE. Smile. Act nice, dance a bit, string them along, you know.

TESSIE. And then?

MAIRE. And then they'll come along and have a word.

TESSIE. What are they after?

MAIRE. Inside information.

TESSIE. They'll never get a thing from them.

MAIRE. Leave that to the boys.

TESSIE. What'll they do to them?

MAIRE. Hurt them a bit maybe. That's all.

TESSIE. Are you sure?

MAIRE. Sure I'm sure.

TESSIE. I don't like to think of him hurt.

MAIRE. Look, you can help or you cannot, it's up to you. Only Kerry's a bit of a frightening fellow, I shouldn't care to argue with him myself.

TESSIE. Oh, God, what'll I do?

MAIRE. Do what you're asked to. Tessie, it's just a knees-up, we've to get them here and entertain them until the lads arrive. Nothing more. What happens after isn't our affair.

TESSIE. Just the sergeant and the other two, leave Tom out of it, why not?

MAIRE. Him most of all. Remember Mrs Cromerty. And Brendan. And God knows how many more of us.

TESSIE. I can't. I can't.

MAIRE. Soldier lover.

TESSIE. Oh, Maire, please. Not Tom.

MAIRE. Tessie. It gets you off the hook. Kerry's a stickler for discipline. He's a loner — they don't approve of him, he can be a bit ferocious, it could be quite a bloody hook. Everything will work out fine — I'll be in for sure and you'll be right again. A clean slate. All forgotten.

TESSIE. Maire? Not Tom.

MAIRE. All of them. Your squaddie too.

TESSIE. Sweet Jesus, what'll I do?

MAIRE. You can't be in a war with your arms folded. You'll have to choose.

The park at dusk.

TOM. I'd rather be alone with you than at this knees-up.

TESSIE. Only for a while, then we could slip away.

TOM. OK then.

TESSIE. And the others?

TOM. They'll leap at it. Booze and birds. No chance of them not coming.

TESSIE. Nine o'clock then. Don't lose that address.

TOM. I won't. I won't.

TESSIE. You're sure we can —

TOM. I've told you how we fiddle it twice. We'll make it. Barring any sudden aggro.

TESSIE. That's great then. I must go.

TOM. You're really cunning, aren't you, Tess? You've set a trap for me, haven't you?

TESSIE. What?

TOM. Lured me with your flashing eyes and your little boy's knees, that's what you've done. And I've fell right smack in. I've had it, haven't I?

TESSIE. Tom, I —

TOM. I've bloody had it. I'm in love with you. Never thought I'd hear myself say that, not in cold blood sort of thing. Next thing I'll be asking you to marry me.

TESSIE. Oh, Tom, it's no good for us, no good at all.

TOM. It's bloody marvellous, I'm not used to it, that's all. Hey, Tess, you're not supposed to cry, I'm telling you I love you, I thought you might be pleased.

TESSIE *sobs.*

You're not pregnant, are you? Because if you are . . .

TESSIE. No.

TOM. I wouldn't mind, you know. I meant it.

TESSIE. Oh, you're so stupid. No, I'm not pregnant.

TOM. Come here. Let's have a hold of you.

TESSIE. You're very strong, aren't you?

TOM. Fit enough. Why?

TESSIE (*in a whisper*). It's easy to die here.

TOM. Is that why you're so moody? Is it? Listen, there's more of us killed on NATO exercise than over here. I'll be OK love. I finish my time here in four weeks, three days, ten hours, ten minutes and sixteen seconds. And a lad copped it from Keighley a couple of weeks ago, so that should see me out. Two of us from the same place in less than six weeks — never. Like lightning. Doesn't happen twice. Anyway, I've got my rabbit's foot. OK?

TESSIE. OK.

TOM. You need some sunshine in you. I'll take you home, soon as I can, take you to the moors.

TESSIE. Don't you understand yet? We couldn't go anywhere if we wanted to, there's nowhere for us to go. You're talking fairy tales, no one'll let us. Tom, do you believe in God?

TOM. There must be someone up there looking after us, guiding us. Tripping us up. Gobbing on us from a great height.

TESSIE. I wish you did believe, I really wish you did.

TOM. I've got this prayer in my wallet, I give it a go sometimes. St Joseph's prayer. He that says this prayer twenty times on successive days —

TESSIE. Shall live life eternal.

TOM. How do you know that?

TESSIE. It's one of ours.

TOM. Get away. There you are, then.

TESSIE. Never therefore forsake me, and whatsoever grace you see most necessary and profitable to me, obtain it for me now and at the hour of my death.

TOM. There you go, you can say it for me. Tell you what, you look up in the bright blue heavens, find God and ask Him for me what's His game. See you at nine. And you be careful.

TESSIE (*in a whisper*). Take care yourself.

TOM. Hey! Come back!

She does.

You've got eye stuff under your eyes.

TESSIE. Oh God.

TOM. It's not a major tragedy, just a mess. Hold still. (*He spits into a handkerchief.*) Hold *still*. That's better.

TESSIE. Is it?

TOM. You need looking after, don't you? Look after each other, should we? Tess? I don't know what you mean when you do that, shake your head like that. I love you, though. Remember that.

TESSIE. Will you say it tonight? Before we meet, before the party?

TOM. I'll say it anytime you want. I love you.

TESSIE. The prayer, Tom, the prayer.

TOM. Oh yeah, sure. Go on, get on home, put your fancy gear on for tonight. Roll on nine o' clock, then, eh?

TESSIE. I left him. Barbed wire squeezing round my heart, the wind in my throat. I chose. Oh, God forgive me. I chose for Brendan, broken by fist and hood and hard white light. For Mrs Cromerty, the life almost blown out of her beneath her cloth. For Maggie Ryan, slit on the mouth and then between the legs. For me. For me. I chose in the park, our park, there in the snow, under the pale communion wafer of the sun. I lost my soul. I left him. Tom, with his bony knuckles and his rabbit's foot and his way of saying 'Tess'. I looked back once and he was stood there watching me. Blowing on his cupped hands, changing feet, stamping. The soldiers stamp. When they slam their way up our streets, sparks fly from their boots struck off the pavements. . . I couldn't wave to him. I stood at the dark edge of the park and I looked back. My boots had made a trail of heel marks bitten into the snow, hard black holes. Like bullet holes. A trail that led straight back to him.

TESSIE's *room at night.*

MAIRE. Where the hell have you been? The timing's vital. I've been frantic waiting.

TESSIE. Walking around.

MAIRE. There's no time to moon. It's all laid on, clothes — come on, get out of those and into these — food and drink, lots of drink. A car for after.

TESSIE. After?

MAIRE. You didn't think we'd hang around, did you?

TESSIE. You never said we —

MAIRE. We'll be off out of it. Where's your underwear? There'll be a house arranged for a week or two till everything blows over. I've squared it with your ma, told her when she let me in we're going to the sea. Surprise jaunt. She's very pleased. Look at that, isn't it gorgeous — white'll suit you. Come on, move.

TESSIE. What'll they do?

MAIRE. The less you know the better, just do what you have to and keep quiet. Bra.

TESSIE. Maire, what'll they do?

MAIRE. Hurt them a bit — I told you. Don't think about it. Tights.

TESSIE. Wait. Tell me why. Make me understand. Tell me what I'm suffering for.

MAIRE. You're suffering? Remember the famine. My great-great-grandmother clung to the doorpost while British soldiers tore down the walls of her home round her babies' heads. My great-grandmother, and yours too I dare say, hunted out of ditches, living on blackberries and cabbage leaves and nothing but a stone to sit on. All those rooms crammed with dead people half eaten by rats, piles of them covered with rocks and cloaks and living souls too weak to crawl away. Come on, slap on the war paint. I pinched some of your eyeshadow. Remember our history, Tessie, and forget the squaddies.

TESSIE. Sweet Jesus, how can I?

MAIRE. Easy. I think of my brother in the Kesh. Stuck in a cell with nothing but a bible and a bucket and a warden to scrub his balls for him with a wire brush. All because he has a dream, like me. I think of my mother. Gulping down the Valium — six a day she's on now, they're her consolation. To forget two of her sons, two of them, Tessie. The place is like a morgue, all shrouded photographs and sighs. Do something to your face for God's sake. Try my lipstick, that one's murder on you. All those men and women in our streets, no jobs, no fault of theirs, not a sound tooth in their heads or a clock to tell the time by, nothing to do but gawp at telly and watch the lives and deaths go by. I think of how we don't belong. Our own land, and we don't belong, not anywhere, and the years bleed on and on and nothing changes. Nothing. We're the crack in the cup to the Brits, shit on their boots, the sour behind yesterday's teeth. Mascara. Rouge. And all the time it's they that are the scum on my land, the land aches with the weight of them. Your fancy boots.

TESSIE. No. Not them.

MAIRE. Get them on.

TESSIE. Maire, I'm feared for you.

MAIRE. I feel great. At last I'm doing something.

TESSIE. I feel next to death.

MAIRE. Tessie, we have to try to make things better. It's our duty, it's dishonourable not to, do you not see? We've nothing to lose, because we've nothing. It doesn't matter what stands in the way, or who. What matters is doing what we've got to do. Kick through the bricked-up windows and let in the light. Now the dress. God, you're a rag doll aren't you? Come on, lift your arms up.

TESSIE. Like surrender.

A long zip is done up.

MAIRE. You look gorgeous. (*She squirts a scent spray.*) Ready for anything. The white is perfect, Tess.

TESSIE. Don't call me Tess. Don't ever call me Tess.

MAIRE. It's a rough world, Tessie. You have to do rough things.

There is a tremendous surge of disco music and the drums are deafening. Gun shots, then silence.

WIFE. In a cold-blooded ambush just before midnight three British soldiers, a sergeant, a corporal and a private were shot dead by unknown assailants and one severely wounded. They were off duty and in plain clothes. Their bullet-ridden bodies were discovered in the early hours of the morning. It is believed they had been lured by two girls to a dance hall where their killers awaited them. The murderers and one of the girls escaped, the wounded soldier and the other girl, in a state of hysterical paralysis, are both in hospital. No one has claimed responsibility and the murders have been condemned on all sides.

TOM. Not yet. I'm not ready to die yet. Never again curving a free kick round the wall and in, steak and kidney pud with gravy, pounding out of the mist in cloggy boots . . . used to like the way it rained here, soft, in veils . . . can't see through . . . Mum? Tess? Little boy's knees . . . feel the life in her, the life . . . never . . . the silver bugle's calling over rooftops. On death they follow the book. Strip the bed, empty the green metal cupboard, the locker, pack your kit, count your money . . . Melcron shield, respirator, tin hat, flak jacket, fags, combat jacket, webbing, rabbit's foot . . .

TESSIE. . . . . Bra, tights, pants, lipstick, mascara, scent, boots, white dress. My white dress drenched in blood. It came out of his eyes. I was ready for blood, but not from his eyes. Blind with it. Broken. All broken. Dissolved, shapeless with fear, the two of us. I never thought — never dreamed — they did it, I did it, what does it matter which? I chose. Now it's a wilderness, the wind's in my throat and I've lost my soul.

MAIRE. I'm a peace-keeper, fighting for peace, killing for peace. Dear God in Heaven, I couldn't do it if I didn't love our people so. Want drives me, want and rage, I'm gashed with hatred. They cannot quench me. We wait our chance in the dark. We *must* be free. I may not see it in my lifetime, some of us are born to die, but whatever happens there'll be more of us, we'll keep on coming 'til there's enough of them dead to fill hell. 'Til my last breath, stripped down bit by bit, 'til I'm hard and spare as a gun, I stand upon my aching land and damn them. Damn them all.

CHARLIE. A good soldier should be smart in his bearing, be able to give orders, show his enthusiasm in everything he does, and understand the military way of life . . . his wife . . . his life . . . been a grand day . . . grand day . . . my legs, the legs, they don't move sort of thing. No more. Never. And this, my hand, and the head, mine, this head, a hole. It's not whole.

CHILDREN (*whispered*). Now we've only got one arm, now we've only got one leg, now we've only got one eye, now we've only got one lung.

WIFE. I knew it was him. Because they told me. I couldn't tell from anything I say — it wasn't him at all. First sight of him I fainted. They had six doctors working on him night and day with their knives and lights and probing fingers in the broken eggshell of his head, trying to keep the life in him. It's like his head blew up.

MOTHER. She's lost to me, I know that. Lost as if she was dead. Except she's not. If I could stand by a cross on a piece of ground and know she was under it I think I might feel better, I could have a priest to her, God have mercy on her soul, forgive her her sins. Ireland's one long funeral. In my father's time they killed, but his was a decent war, they fought it clean, they'd never have done a thing like that, never in this world. Sometimes I feel a blind, sickening rage. What chance did she have in her life, what chance has she got now, stuck in that place they've got her, bent forward in her chair holding her stomach like it was full of poison, watching the blank white wall, silent, struck with a fearful ice? She was the light in this house, the heart, the strength, and now she's gone.

WIFE. Why Charlie? What had he done wrong? He hadn't done anything, only his job, only what he's paid to do, defend us, keep the peace. Oh, Charlie, I do miss you. I miss the chat. Sometimes he shouts and you can't tell what he's shouting. You can tell it's something dreadful. Mostly he's pretty cheerful, though. They make a great fuss of him, the nurses. Last night I dreamed he came back, he walked in at that door and he said, 'Come over here and have a cuddle'. Then he swore and kicked the chair, 'When are we going to get a decent bloody chair?' he said. He always kicked that chair, because army chairs have these great wooden arms, you can't get close together. It was so real, so sort of ordinary, you know, when I woke up I was laughing away —

MOTHER. My house is marked now. They've had me out two, three times in a week, then they'll leave it a week or two, and then again. I've told them, I've been to see her in that place, she's a stone, there's no one left inside her now at all, she's not worth putting on anyone's death list. There's no use kicking my door down, I said, there's nothing here for you. Nor me neither.

WIFE. I know he loves me, even though — I know that for absolute certain. It keeps me going knowing that, that and the tablets. It'll get better all the time. I'm coping — they're very keen on you coping. I haven't cried today at all. I keep busy, the baby sees to that, Morton Charles Riley, and I wait. He's got gadgets in him now. They say he'll improve, course they can't give him a guarantee, they can't tell how he'll be in the end. Today he got his hand up to his chest. On purpose. And he knows me. And the baby. We have a laugh. Calling him bionic, and that.

MOTHER. Silence for weeks and months, all the life burnt out of her. And then the screams. For seven days she screamed like a trapped bird, they let me visit, in case I could reach her. And then today — oh my Tessie, today she smiled at me while tears dropped quietly down her lost young face, she clutched my hand. Will I be dead and gone before they let her go?

WIFE. We haven't talked about it because — well, he can't talk really. Words, but not talk. But I know Charlie. He's such a loving man, he'd say if he could, well, that's it. That's done. And he's right. It was hard at first trying not to hate them, but now — see, you can't love and hate at the same time, at least I can't. They say you can in books but I don't know how. And I love him, my Charlie.

MOTHER. Someone'll have to help us soon.

WIFE. So it's sort of God forgive us all and I hope they're sorry after.

CHARLIE. Feel bitter, no, better now. Hoping. Hoping I'm going to be back with my, with my mates. Re-join up sort of thing. When my legs, my head sorted out. Sorting me out now. I'm chuffed. Yeah. Miss my mates. Once a dole — dole — no, soldier, always. Army my life.

TOM. My life.

TOM
CHARLIE $\}$ (*spoken*). Are you ready for a fight? For we are the English —

MAIRE
TESSIE $\}$ (*spoken*). Yes, we're ready for a fight. For we are the Roman soldiers.

WIFE
MOTHER $\}$ (*very quietly*). Bang. Shot. Fire.

Silence.

THE ANGELS THEY GROW LONELY

by Gerry Jones

Gerry Jones was born in Colwyn Bay, North Wales. He began his professional career as an actor in 1957. Whilst a member of the Old Vic Company in 1962 he had his first radio play accepted and in the next few years wrote many radio and television plays until joining the BBC Script Unit in 1966.

For the past sixteen years he has been a director for the Radio Drama Department and still writes the occasional play. In 1976 he won Spain's International 'Ondas' award for his play *Snake*, which like *The Angels They Grow Lonely* was directed by Martin Jenkins. His novel *The Sin Eater* was published in 1973, and he has directed two of his own plays in the theatre.

He lives in Twickenham with his wife and two children, both of whom are attending drama school.

The Angels They Grow Lonely was first broadcast on BBC Radio 4 on 5 March 1983. The cast was as follows:

NARRATOR	Jim Norton
GEOFFREY JOHNSON	Nigel Anthony
DR CONWAY	Stephen Thorne
HANNAH	Jean Trend
DR WILLIAMS	David Gooderson
MR BLAKE	Robert Lang
AMBULANCEMAN } MALE VOICE }	Roger Walker
RECEPTIONIST } FEMALE VOICE }	Hilda Schroder
THIRD DOCTOR	Edward Cast

Director: Martin Jenkins

When Marilyn Monroe's ability to function collapsed under the weight of stress and sedation, she was taken to a New York clinic concerned with mental and nervous disorders. Marilyn was not told she was taken to a section that housed the mentally ill, and as she was led along a corridor she looked about her with wild eyes.

'What are you doing to me?' she asked. 'What kind of a place is this?'

They just smiled at her and led her on.

NARRATOR. It was eight o'clock on a July morning when Geoffrey
Johnson saw his life begin to fall apart. He felt himself step away
from things as the right side of the room went out of focus and a
bright star began to throb with light deep in his head.

People in white drifted in and out of his shadowy days until
September came and he returned home from hospital. As Johnson
convalesced in the last of the summer sun, he knew things would
never be the same. He felt an invisible straight-jacket about him, a
separation from reality that was difficult to define.

It was in October that the dreams started, or to be more precise,
the dream. He'd never had this dream before but now it came almost
nightly, clearer than any dream he'd ever known. So vivid he felt it
was actually happening to him.

It was November when the dream *became* reality, when Geoffrey
Johnson found the dream and the reality one.

A heavy roll of thunder sounds.

He was sitting alone, watching a television programme on the life of
Al Jolson. The picture was cracking and jumping because of a
violent electric storm outside. The room itself seemed a magnet to
electricity. On the wall a painting of a clown draped in autumn
leaves appeared to be alive with light.

Another crash of thunder and a near and frightening stab of lightning.

TV. It isn't raining rain you know,
    It's raining violets.
    And when you see clouds
    Up on the hills . . .

Thunder crashes again.
Jolson's voice distorts into an unrecognisable sound.

NARRATOR. As the storm burst around the house and the television picture faded to a white square, Geoffrey Johnson stretched his arms out into the electrical current filling the room and felt his body rising upwards. The pull of gravity fell away and he floated upwards, weightless, to the pale cream ceiling. With a gentle push against the ceiling he floated back to the floor. Sense in the normal way no longer had meaning. He knew that at any time he wished he could defy accepted laws and float in space like a bird.

Another flash of lightning.

He felt alone and frightened, a solitary man on the moon, surrounded by alien threat.

A doctor's surgery.

JOHNSON. Dr Conway, can I speak to you in the utmost privacy?

DOCTOR. Of course, of course.

JOHNSON. Well . . . well . . .

DOCTOR. Yes?

JOHNSON. Well . . .

DOCTOR. What is it, Mr Johnson?

JOHNSON. Well — I'm a bit embarrassed about this actually, it's not an easy thing to talk about.

DOCTOR. Oh, I see. Well, don't worry, Mr Johnson, sex after all is a joyous thing, a God-given gift, nothing to be ashamed of — just take your trous —

JOHNSON. It's not sex.

DOCTOR. Pity. (*Pause.*) I'm really rather good with patients who have sex problems. Well now, what exactly is the trouble?

JOHNSON. I've started to fly.

Pause. The DOCTOR coughs. An ambulance passes in the distance.

DOCTOR. You mean yourself?

JOHNSON. Yes.

DOCTOR. You've started to fly on your own?

JOHNSON. Yes.

DOCTOR. Pity . . . I'm not so good with patients who fly. Not what I was trained for really.

JOHNSON. What am I going to do?

DOCTOR. You're going to keep quiet about it, Mr Johnson, that's what you're going to do.

JOHNSON. I have been. No one but my wife knows.

DOCTOR. Good. They don't like it, you see.

JOHNSON. They? Who are they?

DOCTOR. I'm not in a position to say. I can only tell you that it is a condition not fully understood, and certainly not liked.

JOHNSON. But I can't help being the way I am.

DOCTOR. Indeed not.

JOHNSON. Why is it not liked?

DOCTOR. Best not to talk about these things.

JOHNSON. You still haven't said what I'm to do.

Pause. An ambulance passes near by.

DOCTOR. Look, I'll give you a note so that you can see a doctor who specialises in the treatment of 'angels' as you people are known.

JOHNSON. You mean there are others?

DOCTOR. Indeed yes.

JOHNSON. How many?

DOCTOR. God knows. The state of things being as they are at the moment they naturally keep a very low profile about their condition. Anyway, the specialist will tell you all you need to know.

JOHNSON. Thank you, Dr Conway.

DOCTOR. Not at all. Just remember to keep it to yourself. Your condition puts you in some danger.

JOHNSON. I'll remember.

DOCTOR. And remember this . . . have you ever seen anyone flying?

JOHNSON. No.

DOCTOR. Exactly, Mr Johnson. Exactly.

Outside an ambulance approaches nearer and nearer, its blaring horn becoming deafening. It cuts on the sound of a teacup returning to a saucer. JOHNSON's *wife speaks.*

HANNAH. What did the doctor say?

JOHNSON. He took it very calmly really. I was amazed.

HANNAH. Are you sure you explained it all properly?

JOHNSON. Of course.

HANNAH. You told him you've started to fly?

JOHNSON. Yes, yes.

HANNAH. And he took your word for it?

JOHNSON. Yes.

HANNAH. Well, you're not the only one who's amazed. Didn't he ask for a demonstration?

JOHNSON. No.

HANNAH. Incredible. Really, I find it incredible.

JOHNSON. Well, it's not a thing I'm likely to lie about, is it?

HANNAH. I would still have expected him to ask for a demonstration.

JOHNSON. Hannah, if a man goes in with a very bad cold the doctor doesn't ask him to sneeze, does he?

HANNAH. You haven't got a very bad cold — you've started to fly.

JOHNSON. Well, the fact of the matter is, it appears I'm not unique.

HANNAH. There are others? Others who do it?

JOHNSON. So it seems.

HANNAH. How many?

JOHNSON. It's not known. However the doctor made it very clear to me that I've joined a threatened species.

HANNAH. Threatened in what way?

JOHNSON. I'm not sure.

HANNAH. Is there a cure?

JOHNSON. I'm not sure about that either, all I know is I've got to see a specialist.

HANNAH. A specialist in flying?

JOHNSON. Something like that.

HANNAH. Oh, Geoffrey — whether we paint the bedroom green or pink suddenly seems so unimportant.

JOHNSON. Well, what is important is that you don't breathe a word of this to anyone.

HANNAH. Why?

JOHNSON. It's very important, Hannah. I could be in bad trouble.

HANNAH. What kind of trouble?

JOHNSON. It's best not to talk about it, that's all.

HANNAH. Oh, certainly. I wake up in the morning to find my husband floating over the wardrobe and I'm not supposed to talk about it.

JOHNSON. If they got to hear about it they could make things very bad for me.

HANNAH. Who are you talking about?

JOHNSON. Please, Hannah, people may be listening.

HANNAH. I'm listening — and I think I deserve an answer.

JOHNSON. It's just that ever since this started I've had a feeling of my safety being threatened, my liberty; and the doctor today seemed to confirm that feeling.

HANNAH. Well, I can't say I understand.

JOHNSON. I can't say I can either.

HANNAH. Geoffrey . . . what is it like — flying?

JOHNSON. It's difficult to explain.

HANNAH. Is it nice? Is it exciting?

JOHNSON. In a way it is, but it's frightening, too. I'm always glad when I get my feet firmly back on the ground again. Sometimes I'll be enjoying the sensation, losing myself in it, then suddenly I'll be terrified that I'll just float and never get back. I feel that I'll end up suspended in a sort of eternity of imprisonment, unable to speak, praying for death of some kind.

The phone rings. JOHNSON *picks it up.*

Hello?

VOICE. Mr Johnson?

JOHNSON. Yes.

VOICE. Welcome to the Angels, Mr Johnson.

JOHNSON. Who is this?

VOICE. It's unimportant, just a friend. I'm phoning to warn you of the ambulances.

JOHNSON. Which ambulances?

VOICE. You'll know them when they come. Take care.

The phone is put down at the other end.

HANNAH. Who was it?

JOHNSON. Someone warning me about ambulances.

HANNAH. I beg your pardon?

JOHNSON. God, I'm going crackers. Let's hope the specialist can help.

HANNAH. What's his name, by the way?

An office. A door opens.

JOHNSON. Dr Williams?

DOCTOR. Yes, come in. Come in, Mr Johnson.

JOHNSON. Thank you.

DOCTOR. Shut the door, please.

> *The door is shut.*

> Sit down, please.

JOHNSON. Thank you.

DOCTOR. Now I've been looking through your medical file and it's quite hefty, isn't it?

JOHNSON. Yes, it is. I've had quite a bit of ill health in the last few years.

DOCTOR. Yes. (*He turns several pages over, then stops.*) Correct me if I'm wrong, but for several years you suffered with high blood pressure of a quite acute nature and you were treated for this with various drugs.

JOHNSON. That's right.

DOCTOR. However, you gave up taking the prescribed treatment of your own accord and you later suffered a cerebral haemorrhage.

JOHNSON. Yes.

DOCTOR. After the cerebral haemorrhage you were left with scar tissue on the brain. This was detected in a brain scan given to ascertain the cause of headaches and double vision. It now appears that the scar tissue has led to further unexpected and undesirable side effects — to wit, a tendency to gravitational disorder leading to balance instability elevationally speaking; or to put it more precisely — flying.

JOHNSON. Yes.

> *Pause.*

DOCTOR. The condition in which you find yourself, Mr Johnson, while not, as you know, unique, is indeed a most regrettable one. Your body's determination to disobey the accepted laws of nature make this a matter with problems over and above those of a medical nature. Are you in any way aware of the next stage of your development?

JOHNSON. No.

DOCTOR. Perhaps it's just as well.

JOHNSON. What on earth do you mean?

DOCTOR. Well Icarus, after all, perished when flying too close to the sun, did he not?

JOHNSON. Perished? You mean I'm going to die?

DOCTOR. No, Mr Johnson, it's not a case that involves fatality. You won't die.

JOHNSON. Perish but not die?

DOCTOR. Now with regard . . .

JOHNSON. Perish but not die?

We hear JOHNSON *on echo saying:*

I feel that I'll end up suspended in a sort of eternity of imprisonment, unable to speak, praying for death of some kind.

An ambulance rushes by outside, the horn blaring. Pause.

DOCTOR. What's the matter, Mr Johnson?

JOHNSON. I'm all right . . . it was just . . . er . . . nothing.

DOCTOR. Tell me, can you control these impulses to fly?

JOHNSON. It's becoming more and more difficult.

DOCTOR. Hmm.

JOHNSON. What can you suggest for me?

DOCTOR. Well, of course there are certain tablets, tablets which if taken in sufficient number could possibly restrict your aerial tendency.

JOHNSON. Oh, good.

DOCTOR. They would also, unfortunately, restrict any tendency you may have to go on living — although I must point out to you that in your present circumstances that may be no . . . I see that you are married.

JOHNSON. Yes, ten years.

DOCTOR. How is your wife coping with this situation?

JOHNSON. I don't think she's quite taken it in yet.

DOCTOR. In your file it says you have no children. Is that correct?

JOHNSON. No, we have a small boy, now. He was born six months ago.

DOCTOR. Thank you, I'll correct the file. I must also inform you that it is possible your child has inherited your 'tendency'.

JOHNSON. Oh, no.

DOCTOR. We usually find this condition to be an inherited one.

JOHNSON. But there was nothing wrong with my parents.

DOCTOR. Indeed not, and that is where your child may be lucky. Your condition was brought about through illness, so there is a strong possibility that your son will not have inherited the gene. Nevertheless, Mr Johnson, I would like to have him brought in so that he can be treated.

JOHNSON. In what way?

DOCTOR (*Changing subject*). In your place of work, what do you do exactly?

JOHNSON. Designer, graphic designer.

DOCTOR. Do you like your job?

JOHNSON. It's all right. As a matter of fact I was given promotion some time ago, that makes it a lot better.

DOCTOR. Does your boss know your medical history?

JOHNSON. Some of it.

DOCTOR. But nothing of this latest development?

JOHNSON. No.

DOCTOR. Well now, I'm going to arrange for you to have a series of tests. You will have to come into hospital for a few days.

JOHNSON. Do I have to undergo tests?

DOCTOR. I am afraid it is essential if we are to know how to deal with this matter.

JOHNSON. But I can't take time off work — I'd have to give my boss a reason.

DOCTOR. All that can be arranged — don't worry.

JOHNSON. I am worried.

DOCTOR. You must look at it from my point of view. At the moment, in spite of everything you've told me, I only have your word for it that you have this condition. I need proof.

JOHNSON. Why would I make up such a story?

DOCTOR. You'd be surprised what people imagine of themselves.

JOHNSON. Dr Conway didn't ask for proof.

DOCTOR. Dr Conway's job is simply to pass people like you on to people like me.

JOHNSON. I feel afraid.

DOCTOR. Of what?

JOHNSON. I don't know.

DOCTOR. There's no need for fear, provided you're honest and give us the proof we need.

JOHNSON. I am being honest.

DOCTOR. Proof, Mr Johnson. That's why you must have tests. Unless of course you can give the proof I need here in the office. Proof that you can fly.

JOHNSON. Well . . . I could . . .

DOCTOR. And will you?

JOHNSON. Well, yes . . . yes, all right.

DOCTOR. Good. That's excellent.

JOHNSON. One thing, Doctor. You said that once you had proof you would know how to deal with the matter. How exactly would you deal with it?

DOCTOR. In your best interests, of course.

JOHNSON. I'd like to know what my best interests are.

DOCTOR. In time. Now would you please be good enough to demonstrate to me your ability to fly.

The intercom on the desk buzzes.

DOCTOR. Yes?

VOICE. I'm sorry to disturb you, Dr Williams, but you're needed in 24 right away.

DOCTOR. But I'm with a patient at the moment, this is most inconvenient.

VOICE. I realise that, Doctor, but this is urgent, I'm afraid. Category A.

DOCTOR. Oh, very well, I'll be there as soon as I can. (*He clicks off the intercom.*) Damn! I'm sorry, I shall have to cut short our meeting, I do apologise. Perhaps you'd be kind enough to have a word with my secretary and make an appointment to see me as soon as possible.

JOHNSON. Right. Thank you, doctor.

DOCTOR. Goodbye, Mr Johnson. I'm really most sorry.

JOHNSON (*moving to the door*). Goodbye, Doctor. (*He opens the door.*) Oh, by the way — once you have your proof — is there a cure?

DOCTOR. Yes, there's a cure.

JOHNSON. Well, that's something.

DOCTOR. Goodbye, Mr Johnson.

The door closes. The DOCTOR *dials a digital number.*

Hello, Dr Williams here, I want to talk to Control, please.

NARRATOR. On a sudden impulse Johnson walked past the secretary without even glancing her way. He left the hospital conscious of a feeling of escape, or certainly the need for it. He mingled with the home-going people who had just wiped away another of their days. People protected by the numbing security of the music of the clock, time was what made them tick. How he envied them as down the pavements of the city their homeward tide broke off in fragments. They could disappear into similar houses where their wives and children waited to absorb them into the day's predictable conclusion. He walked among them, yet apart. They, for all their hundreds or thousands or millions were no more than a grain of sand in the Sahara, but he felt, for all his physical anonymity, like a golden eagle in a spotlit cave. He felt instinctively pleased that he hadn't demonstrated his flying ability in the office, and was determined from that moment to say nothing.

An ambulance approaches.

He doesn't know that his nightmare is now in the computer and that the betrayal must have its end.

The ambulance comes nearer, its siren blaring. Suddenly the sound cuts dead.

HANNAH. . . . ime it is?

JOHNSON. What?

HANNAH. I said, do you know what time it is?

JOHNSON. Oh . . . time? Er . . .

HANNAH. It's gone midnight. You said you were coming to bed two hours ago.

JOHNSON. I'm sorry.

HANNAH. I was fast asleep. I woke up and you weren't there. What have you been doing down here?

JOHNSON. Oh, this and that.

HANNAH. What?

JOHNSON. Reading. I did some reading, I must have dropped off.

HANNAH. You haven't been reading, you've just sat here like a zombie.

JOHNSON. Look, I'll come to bed soon.

HANNAH. You've just been sitting here staring into space, haven't you? You've got to snap out of it.

JOHNSON. Hannah, please . . .

HANNAH. You're like someone living in another world.

JOHNSON. That's how I feel. What the hell is going to happen to me?

HANNAH. Nothing will happen to you.

JOHNSON. It will, something bad. I know it.

HANNAH. Something bad will only happen if you let yourself get into a state like this.

JOHNSON. I just have this constant feeling of threat.

HANNAH. Nobody is threatening you, it's all in your mind.

JOHNSON. You're wrong.

HANNAH. Why on earth should anyone want to harm you?

JOHNSON. We shouldn't be talking like this, we shouldn't even be thinking about it.

HANNAH. Geoffrey, if you go on like this you'll go mad.

JOHNSON. I don't want to talk about it.

HANNAH. We've got to talk about it. The hospital phoned me this morning.

JOHNSON. Why didn't you tell me?

HANNAH. Because I know that if I so much as mention the hospital it upsets you.

JOHNSON. What did they want?

HANNAH. They were simply phoning to say that they would appreciate it if I would encourage you to take the tests immediately.

JOHNSON. I bet they would.

HANNAH. They say the tests are very simple and that they really are the only way to help you with the problem. Please think carefully about it, it's for the best, really.

JOHNSON. Don't let them fool you, Hannah.

HANNAH. Nobody's fooling anyone. Now, come to bed and have a good night's sleep.

JOHNSON. No, I'll stay up for a while.

HANNAH. Please — it's not good for you sitting here.

JOHNSON. I'll come to bed soon. I'm just going to play some music.

HANNAH. Well, don't be long.

JOHNSON. I won't.

HANNAH. I don't like you sitting down here.

JOHNSON. I won't be long.

HANNAH. All right . . . all right.

JOHNSON. I'm sorry . . . I know it's hard for you.

HANNAH. Just don't be long.

She goes out, closing the door. Music.

NARRATOR. After he'd turned on the music he sat looking out through the French windows at the distant stars. Stars which throbbed across the light years to deep in his head. The room was a kaleidoscope, his mind just pieces of disconnected memory.

The music filled him with its soothing sound and slowly he surrendered his body again to the weightlessness and the beauty of flight. On feather feet he moved out into the garden and with his eyes fixed out into the universe he separated from earth. Bathed in moonlight he began his night flight. He soared and looped above the winter apple trees, over the old air-raid shelter and through the icy air up to the television aerial that seemed to sing with silver. The lonely bird of madness dived and swooped until at last the grass made a prisoner of his feet.

The music is clicked off.

He turned off the music, turned out the light and began to walk up the stairs, not knowing that in the shadowy street outside an ambulance was parked, and that the driver had observed and contacted Control.

MR BLAKE's *office. A knock at the door.*

BLAKE. Enter.

The door opens.

JOHNSON. You wanted to see me, sir?

BLAKE. Yes, indeed — come on in. Sit down, Geoffrey, I just want to finish this. (*He writes for a moment, then stops.*) There. Tell me, Geoffrey, how are you feeling?

JOHNSON. I'm feeling fine, sir.

BLAKE. Are you? Good.

JOHNSON. Is something wrong, sir?

BLAKE. Well, I think we both know the answer to that, don't we?

JOHNSON. Sir?

BLAKE. Geoffrey — the reason I've asked you to come and see me is that I'm having to do a bit of a reshuffle and I thought you'd better be put in the picture. Because of the possibility of a forthcoming merger, it is necessary to rethink some of our existing posts. One of these moves involves your present position, I'm afraid, although of course there's nothing to worry about.

JOHNSON. Which means what, sir?

BLAKE. Which means, purely as a temporary measure, you understand, that you will return to your former post until the new plans have all been formulated.

JOHNSON. I must say, sir, that I've heard nothing of this proposed merger.

BLAKE. Nor had anyone else, Geoffrey, and I must ask you to keep it a secret between ourselves for the moment.

JOHNSON. When do I make this temporary return to my former post?

BLAKE. Well, as there is no point in delaying the new plans, I would suggest that it is in the next few days.

JOHNSON. Who else is involved in this reshuffle?

BLAKE. I'm afraid I'm not in a position to discuss the matter, I wish I were.

JOHNSON. May I ask how long this temporary return is likely to be?

BLAKE. It is pointless at the moment for me to say anything. We shall simply have to wait and see.

Pause.

JOHNSON. You've been 'informed' about me, haven't you?

BLAKE. What do you mean, Geoffrey?

JOHNSON. Nothing.

BLAKE. Tell me what you mean.

JOHNSON. I meant nothing.

BLAKE. What could I have been informed about?

JOHNSON. Nothing, sir . . . really, nothing.

Pause.

BLAKE. I see.

Pause.

JOHNSON. Well, if you'll excuse me, I'll get back to my office and start clearing it up.

BLAKE. Before you do that, Geoffrey, you're needed at the main entrance. I suggest you go now.

JOHNSON. Who is it?

BLAKE. I only know that you were asked to be there. I gather it's important, someone needs to contact you.

JOHNSON. I'll go straight away.

BLAKE. Right . . . and Geoffrey — I'm sorry, but you know how things are.

JOHNSON. I'm sorry, too, sir . . . and I do know how things are.

The office door closes. Cut to exterior and sound of distant traffic. An ambulance approaches, its horn blaring, and screeches to a halt.

AMBULANCE MAN. Mr Johnson?

JOHNSON. That's right.

AMBULANCE MAN. Jump in, sir. I'm taking you to the hospital.

JOHNSON. Go to hell.

AMBULANCE MAN. What?

JOHNSON. I said, go to hell. I'm not getting in any ambulance.

AMBULANCE MAN. It's your son, sir. He's been taken to the hospital.

JOHNSON. By whom?

AMBULANCE MAN. I only know he's been taken there. He's to undergo tests.

JOHNSON. They've got to be stopped, stopped at once.

AMBULANCE MAN. Well, jump in, we'll be there in no time, we can sort things out.

JOHNSON. All right — come on, quickly.

The ambulance drives away. Interior of ambulance driving along.

JOHNSON. If they harm my lad . . .

AMBULANCE MAN. Nobody's going to get harmed, sir.

JOHNSON. It's me they're really after, you know.

AMBULANCE MAN. Is it, sir?

JOHNSON. You know damn well it is.

AMBULANCE MAN. I only know I was asked to pick you up.

JOHNSON. Well I'm going to deny everything, deny it all. I'm going to say I made the whole thing up.

AMBULANCE MAN. Made what up, sir?

JOHNSON. They can't prove a thing.

AMBULANCE MAN. About what, sir?

JOHNSON. Nothing. I'm saying nothing.

AMBULANCE MAN. As you wish.

JOHNSON. I should never have mentioned it in the first place, it was a big mistake.

AMBULANCE MAN. What was, sir?

JOHNSON. Nothing.

AMBULANCE MAN. Soon be there, sir.

Hospital tannoy in the background. Footsteps approaching a reception desk. They stop.

JOHNSON. Could I speak to Dr Williams, please?

RECEPTIONIST. Dr Williams?

JOHNSON. Yes, he's a specialist here.

RECEPTIONIST. I'm sorry, there's no Dr Williams at this hospital —

JOHNSON. Yes, there is. I saw him yesterday.

RECEPTIONIST. I'm afraid not, sir. You're mistaken.

JOHNSON. I see . . . well, could I speak to Dr Conway, then?

RECEPTIONIST. Dr Conway?

JOHNSON. Are you going to tell me there's no Dr Conway either?

RECEPTIONIST. No, you'll find Dr Conway in Room 17 along the corridor there.

JOHNSON (*moving away*). Thank you.

The RECEPTIONIST *picks up a phone.*

RECEPTIONIST. He's on his way to Room 17.

NARRATOR. Johnson walked along the corridor, aware of the fact that several doctors who had been talking together beside the tea and coffee machine had moved away and were now following him closely behind. He walked slowly, terrified of any lifting movement in his feet. He arrived at Room 17, knocked, and was told to enter.

Room 17.

DOCTOR. Come in, Mr Johnson, I've been expecting you.

JOHNSON. And I was expecting Dr Conway.

DOCTOR. Dr Conway? I can't say I've heard of him.

JOHNSON. I don't suppose you can.

DOCTOR. What do you mean?

JOHNSON. Nothing really.

DOCTOR. Are you all right, Mr Johnson?

JOHNSON. Yes, I'm fine — except I'd like to know why my son has been brought here.

DOCTOR. Let me put my cards on the table — it's necessary for you to undergo some tests.

JOHNSON. Why?

DOCTOR. To find out if . . .

JOHNSON. If I'm one of those people who can fly?

DOCTOR. Yes.

JOHNSON. Let me put my cards on the table — I can't.

DOCTOR. That's not true now, is it?

JOHNSON. Perfectly true . . . I was making the whole stupid thing up. I'm very sorry to have wasted so much of everybody's time and I'd be very grateful if we could now forget the whole business.

DOCTOR. I know you can fly, Mr Johnson . . . you were seen.

JOHNSON. If you know I can fly then why do I have to undergo tests to see if I can?

DOCTOR. Cards on the table, Mr Johnson, you're not going to undergo tests — you're going to undergo treatment. If the treatment is successful you won't have anything more to worry about, if not, then we have to make arrangements.

JOHNSON. But I can't fly . . . I can't . . . really I can't.

DOCTOR. I'm afraid it's too late to play silly buggers, Mr Johnson — and there's the question of your child.

JOHNSON. What about my child?

DOCTOR. Let's put it this way — to save any further bother — if you agree to the treatment, then I'm prepared not to give any tests to your child. I should give him tests, believe me, and in normal circumstances I would, in the hope of preventing any trouble in the future. But these tests on a child of that tender age can be very damaging, very dangerous. So, I'll do a deal with you — agree to take treatment and your child can go home at once.

JOHNSON. Why is it so important that I have treatment?

DOCTOR. We have to get things back to normal, don't we?

JOHNSON. Why is it so important?

DOCTOR. You don't want your child damaged, do you?

JOHNSON. This isn't a hospital, it's a madhouse.

DOCTOR. Do we have a deal?

JOHNSON. I'm going to report you . . . I'm going to report you to the authorities.

DOCTOR. Cards on the table, Mr Johnson . . . I am the authorities. Now, just sign this and we'll get started.

Fade out. Fade up on a steady bleeping sound. The bleeping stops. A voice comes through on a speaker.

FEMALE VOICE. We've completed the tests, sir.

MALE VOICE. I see . . . and Mr Johnson?

FEMALE VOICE. Treatment unsuccessful, sir.

MALE VOICE. Very well. Make arrangements.

A loud humming noise. It gets louder and louder — then silence.

DOCTOR. Come on, Mr Johnson . . . Wake up now. It's all over.

There is a crash of lightning. Al Jolson is singing:
'And the angels they were lonely
Took you 'cos they were lonely . . .'

NARRATOR. Johnson stared up at the underside of the ceiling light and saw himself lying in its bulb. For a while he lay stretched out on the carpet, then slowly and awkwardly he got to his feet. The violent electric storm still raged outside. The room itself seemed a magnet to to it. On the wall the painting of the clown draped in autumn leaves appeared to be alive with light.

From the television came an explosion of applause as the Al Jolson programme came to an end. He switched off the set and wearily sat back in his easy chair. This prisoner of a falling sickness knew that twenty minutes had passed and he had been nowhere, seen nothing. Yet if reality and dreams were fast becoming the same, what difference did it make? And what did it matter what names they had? Upstairs his wife and child slept on. At least they hadn't seen him, that was something. No one this time had witnessed his journey.

The storm began to fade away. It had all been a dream of a kind. But then, so was everything in a way, it was getting harder and harder to hang on to some kind of normality.

Mr BLAKE's *office. There is a knock at the door.*

BLAKE. Enter.

The door opens.

JOHNSON. I believe you want to see me, sir?

BLAKE. Ah Geoffrey, dear lad. Come on in, close the door.

The door closes.

BLAKE. You're looking very well, Geoffrey, I must say — better than you've looked for some time.

JOHNSON. Thank you, sir. I'm feeling fine.

BLAKE. Good. Good. Although of course, looks aren't everything, are they?

JOHNSON. Sir?

BLAKE. I mean, they can be deceptive.

JOHNSON. I'm feeling fine, sir. Really.

BLAKE. Take me, for example. I look a pretty fit sort of person, don't I? I'm not overweight, I play a lot of golf, I drink very little and I don't smoke. But I have a secret, Geoffrey. Shortly I shall be retiring early because of ill health.

Pause.

JOHNSON. I'm sorry to hear that, sir.

BLAKE. A bad heart, Geoffrey, would you believe it, a bad heart. So — and this is just between us for the moment, I am being put out to grass.

JOHNSON. You'll be missed, sir. I'm very sorry.

BLAKE. Oh, I won't be missed, not at all. A little reshuffling and things will carry on as normal.

JOHNSON. Have you asked me here to discuss the reshuffle, sir?

BLAKE. Yes, I have, I have.

JOHNSON. I'm very flattered.

BLAKE. When I was first told that I was being put out to grass I felt rather depressed about things, after all, most of my life had been spent here. But then I said to myself: 'Well, perhaps the grass is greener on the other side'. And now I'm sure of it.

JOHNSON. Do I take it that the reshuffle concerns me? My job?

BLAKE. Yes, indeed it does.

JOHNSON. Do you mean that I'm going to get my old job back?

BLAKE. You mustn't jump the gun, Geoffrey, I haven't finished yet. Now when I retire it's going to mean one thing above all the rest. It's going to mean that an enormous feeling of stress will be lifted off my shoulders. We all work under stress here, all of us. The health of everyone who works here is under constant observation. There's not a person here whose medical record is not completely documented and known to me and this detailed information gives me little pleasure.

JOHNSON. Why have you asked me here to discuss your retirement, sir?

BLAKE. I haven't, Geoffrey . . . it's to discuss yours.

JOHNSON. What?

BLAKE. I think you heard me.

JOHNSON. Retirement?

BLAKE. That's what I said.

JOHNSON. But . . . I'm only half way through my career . . . I've . . .

BLAKE. No, Geoffrey, I'm afraid you're at the end of it. Medical reports, you see, medical reports. I only talked about myself and my retirement to show you that you're not alone in having problems, and, as I say, all our problems are documented.

JOHNSON. What problems, sir? What problems?

BLAKE. Oh, come now, dear lad, I know everything about you. The trouble you've been having over the past year or so has been fully reported in spite of your understandable efforts to hide the facts.

JOHNSON. My health has never affected my work — never.

BLAKE. I'm afraid that's a matter of opinion.

JOHNSON. Please, sir . . .

BLAKE. It's certainly not the opinion of our medical adviser. He has presented a clear report showing that you are subject to a high anxiety state. It's in your best interests, you must understand that. Our only concern is for you. Your steadily worsening health is putting you at considerable risk.

JOHNSON. Please, sir . . . please.

BLAKE. It's no good, Geoffrey. The decision has been made.

JOHNSON. Sir . . .

BLAKE. You'll be taken good care of . . . don't worry.

JOHNSON. Help me, sir. There must be something you can do.

BLAKE. I can't even help myself.

JOHNSON. But my work, my family . . . my life, sir, my life.

BLAKE. There's nothing I can do, nothing.

JOHNSON (*shouting out loud*). You bastards, you bloody bastards . . . I . . . Oh God . . . I . . . Oh . . .

BLAKE. What's the matter? Are you having . . .?

JOHNSON. Oh . . . I feel . . .

BLAKE. Sit down, Geoffrey. Take it easy. Come on, now, relax. That's it . . . just relax . . . relax . . . relax . . .

JOHNSON. I feel . . . I feel . . . (*He makes a strange cry:*) Ahhhhhhhhh.

BLAKE. Don't worry, I'll get the nurse here straight away — you'll be all right. Take it easy, now.

JOHNSON *can only make a gurgling sound at the back of his throat.*

BLAKE. The nurse will be here in a minute. Just relax . . . just relax. (*He presses his desk intercom.*) Quick, get the nurse here, Mr Johnson is having a fit. Hurry up, for God's sake, he's going mad.

NARRATOR. People in white. Places of white. Language shuffled beyond recognition and spoken in a world of matching clocks that told a time he couldn't remember. The people in white drift endlessly like ghosts in and out and on as he feels an equally endless loneliness he knows this time will last forever. This time there will be no change, no return. Strange unknown people stare at him and shake their heads. Days and nights became the same, and as he stumbled around the identical and echoing hospital corridors the only thing he could really remember was that he had no memory at all.

When they came to move him to another place he walked on dust through dreams of other times and other places and a woman whose name began with H. Like a prisoner going to the scaffold he entered the waiting ambulance.

He sat slumped in the seat as it moved through the city and out towards the suburbs, on towards somewhere else. He did not notice the changing backcloth of buildings and trees and the long sad rows of shops and empty places where houses used to be and people used to live. He just sat staring at his feet. conscious only of a hopelessness and the vehicle travelling further and further into isolation.

In the fragments of his mind he asked the driver, 'What are you doing to me?' 'What kind of place am I going to?' But no words came and the driver only smiled in the silence.

After a while the ambulance stopped and the driver spoke. For the first time in a long, long while Johnson understood what was being said, the words were sharp and clear and had meaning.

DRIVER. Here we are, sir . . . you're there . . . home.

NARRATOR. It was then that Geoffrey Johnson looked up from his feet and saw his 'home' was a vast birdcage in which screaming people were flying in despair, crashing themselves continuously against the silver bars.

NO EXCEPTIONS

by Steve May

For Ch

Steve May is 30, was born near London and studied in Cambridge and Bristol, where he now lives with his wife and young daughter. His plays include *Down Among the Umalogas*, *Poisoned Apples*, and *Jack*, all for BBC Radio.

No Exceptions was first broadcast on BBC Radio 4 on 15 October 1983 with Rod Beacham as the TEACHER, produced by Alec Reid.

A medium-sized primary school staff room at break. Indecipherable background talk and cup noises, with a little laughter. The conversation we pick up is private, not general.

TEACHER. Roger Burge? How could I forget. What's he done this time? Typical. Dot said he had a thing about cars. You know Dot up at the big school — give her Ghengis Khan and she'd have him down for the school choir. Said our Roger was after a job in a garage — look good on his application form, won't it?

A bell rings loudly. General movement and cup clinking. Teacher also rises and starts moving to the door.

Surprised they caught him, the way he could run.

The door opens into a slightly echoey corridor full of moving children.

I couldn't keep up with him. Gave up trying. Bad for my image, getting beat up by a ten year old. (*He shouts.*) Jamie East walk.

Walking down the corridor.

Had something about him, Roger — the little so-and-so. There's the photo. Hall of fame. When he won the district cross-country — nine, he was, then. Be fifteen now, I should think. Grinning all over his face.

Lose background.

Moody, sulky, that was his trouble. Something goes wrong, off into one of his moods. Wouldn't say a thing. Not a word. Like talking to a brick wall.

On the tarmac outside the school. It is windy. The sound of children's voices. They are chilly, jumping and chattering to keep warm.

TEACHER. Right — who wants to meet my friend?

Slapping of a heavy bat on palm. There is silence, a few giggles.

Brian Jennings? He'd like you, my friend. Especially your backside.

Big slap on palm of hand, followed by laughter.

Right. Today we're going to run. Not like yesterday. Not like a bunch of spasos and grandmas. We're going to run properly. And anyone who doesn't want to run properly can stay in the classroom and do some English.

Some groans.

Chris Coke — anything to say? Hard man Chris — crying like a baby yesterday in football, weren't you?
Weren't you?
Yes, what?
Yes, Sir.
You won't get your backsides hit (*slap of bat*) for coming last, or not running as fast as Roger Burge. Run as fast as Roger Burge and you'll be in the next Olympics. You get your bums warmed up for not trying. For messing about. For chatting. Got it? . . . You know where to go. Out the first gate, over the footbridge, along to the park, round the hut, and back. Five minutes.

There are groans and laughter.

Right, line up. First whistle for the girls, second for the spasos, third for the runners. Patrick are you a runner? Well, then, get in with the spasos. Ready.

The whistle sounds.

Pounding adult feet, follow pounding children's feet at a distance.

(*Breathing a little hard*). There he goes. Through the park gates. No one near him, not even Brian. I can't keep up with him. Too much weight. Mind you, I was a sprinter. No good at distance. Nearly ran for England, once. Reserve. Wouldn't have heard of me. Sixth in the Three A's. Never won anything, see. Got down to ten, three. Wind assisted, that was.

Here he comes. Only happy when he's running or causing trouble. Not even puffing. (*He shouts.*) Go on, Roger, work. Kid like that, makes you think twice about original sin. Not that he's vicious or malicious. No way. Just does things. A doer. On the go the whole time. Devil makes work. Give him a job — carrying gear, moving furniture — happy as a pig in muck. (*He shouts.*) That's good, Patrick. Don't let them get away from you. Ask him to sit still for five minutes — no chance.

Lose background.

Take maths. Just sits there writing numbers. And signs. So slow it's painful. No rhyme or reason to it. Just rows and rows of numbers and plusses and minusses and timeses. Sort of trance, he's in. That's the best of it. Otherwise he'd be breaking up the classroom. I let him get on with it. Limit to what you can do.

In the classroom. It is quiet apart from some coughing and seat noises, and chalk on blackboard.

TEACHER. . . .That's right, and then you copy it down in your exercise books. Neatly, Adrian. And when you've copied out what I've written on the board, you draw a lovely picture, like mine.

There is mild laughter.

What are you laughing at, Julie? You couldn't draw a curtain.

More laughter.

(*Aside*). Used to try and get them to write it on their own, like they teach you at college. You know, self-expression. Half of them never write a thing, and when they do it's painful. Take you all weekend to read it, let alone mark it. Talk about deprivation, inner cities, social background — most of them are just plain thick. No one wants to admit it, though.

(*Brusquely.*) Yes, Bobby? What? A 'c' in exception? Have to 'c' about that, won't we?

Giggles.

Look it up in the library — (*restraining.*) Not now — at break. Just copy what's there for now.

(*Aside.*) Never was any good at spelling.

During the following the noise level rises, but never very loud.

Trouble with me is I'm too soft. I am. I like messing about too much, having a laugh. All one big act, this shouting. Bluff. Thing is, they never call it. Don't ask why. S'pose they're frightened of going too far — over the limit. Like to know where they stand, kids do. Everything thought out for them. Copying off the board . . . they're happy doing that. No thinking involved. All this finding out for themselves, they hate it. Makes them feel insecure. Had their IQ tests the other day. Some bloke came down. Know what the highest score was — (*He shouts.*) Stand up the talkers. Stand up anyone I asked to talk. That's funny. No one. What was the noise, then? Mice? Roger Burge's brain working? Clank, clank, clank.

Giggles.

Sharon, button it. You can talk to your boyfriend at break.

Stifled giggles.

(*Aside.*) See what I mean? Highest was a hundred and seventeen.
Average is a hundred. Only three of mine over a hundred. Couple of
them didn't even score — like Roger Burge. Stop talking. If you've
got something to say say it to me. They start at seventy — that's the
cretin level — and two of mine didn't even get seventy. And this is
the top class. Bill's got the so-called remedials. What can you do?

(*He calls, not very loudly.*) Turn round, Patrick.

(*Aside.*) That's why we have PE every day. Exercise their bodies.
They can understand that — running and jumping and kicking each
other. Mind you, there's some of them don't like that either. Muntaz
Ali stop talking. Thick and lazy. Can't stick that. So competitive
myself. Always have been. Whingers, they drive me up the wall. Have
to go flat out, I do, even when I'm just playing stupid games with
the kids. Way I'm made. If you don't take it serious the whole thing
just falls apart. They sense it. Mind you, some of the kids can run.
Our team's won the district cross-country the last two years. Roger,
first last year. You've seen him in action. Look at him now, all
slouched, fingers in his mouth, staring into space. Bet he hasn't
written a word all morning. Run? I can't keep up with him. Brian
and Chris can beat him in the sprint, but over half a mile, or a mile,
or a mile and a half — you can't stop him. Keeps going all the way.
Last year in the cross-country he beat the record for under-fourteens,
and he was only nine.

(*He calls.*) Roger Burge stop flapping your ears and do some work.
I don't know what you've got to smile about. Burge the Splurge.

Giggles.

When Roger smiles it reminds me of the old song — how does it go —
'When you're smiling,' (*He sings.*) 'the whole world smiles with you.'

Groans and laughter.

Never seen a smile like it. (*Aside.*) On the game, his ma. And he's
a bugger. Always on the brink. Slightest thing sets him off. Go for
anyone if they upset him. Not interested in school. Can't hardly
write his name. Should be in Bill's class.

(*Aloud.*) Let's have a look, then, Patrick.

The rustle of a book as it is handed over.

(*As he marks.*) Not bad. Not bad. No 'v' in mother. What are you
doing Saturday week, then Patrick? Good. And don't go fixing
anything up with your girlfriend. We might need you.

(*Incredulously.*) What's happening next Saturday week? Tell Patrick
what's happening next Saturday week.

(*Echoing their audible reply:*) The district cup. Cross-country. And
who won it last year? We did. And who's going to win it this year?
Us? We might, if Roger Burge pulls his finger out. Keep your hair
on — I never said anything about running. Nothing wrong with his
running, long as he goes flat out and gets a bit of competition, not
like you spasos. It's your behaviour, isn't it Roger, your work in
class. Bring your book up.

(*He shouts.*) Bring your book up. You may be thick but you're not
deaf.

Pause.

Thank you, Roger. (*Flicking pages:*) Very artistic. One, two
three . . . four, five . . . six words. Six words in twenty minutes.
See these lines here — here — across the page. They're for writing on.
Like railway lines, and your pen's the train. Get it?

(*He bellows.*) Now do it again and do it properly.

A light slap of a book on a face.

(*Aside:*) Don't know why I bother sometimes. Trouble is, I like him.
Got something about him. I don't know. And he can run. Pain in the
arse, though. Look at him. It's a picture. Trying to behave.
Concentrating fit to bust; one eye on me all the time. All the energy
tensed up and trapped inside him — like a spring ready to snap. Got
his mind on running — the district — kill him to miss that.

(*He calls:*) Mandy, work.

Sin, really. Silence, sitting still, books. Not his nature. Needs to be
up and about, shouting, throwing things. What can you do? Ease up
on him, let him get away with anything, and they'll all be off. Kill
or be killed, in this business. Give them half a chance and they'll
trample on you.

He stands and begins to circulate the classroom.

Very good, Roger. Much better. That's ten minutes.

Ten minutes you've behaved yourself. Question is, how long you can
keep it up for. Not just ten minutes. Not just this morning. Not just
this afternoon. Every morning, every afternoon. Every day, every
week. You're in my class for a whole year, remember. It's like
running. No good just sprinting when you think I'm watching — like
most of you do. Grabbing a pencil and copying off Donald 'cos he's
too much of a sissy to stop you. You've got to keep at it all the
time, flat out. Because remember, I'm watching you all the time,
every minute. Running for this school is a privilege, not a right.
You've got to earn it. And remember — it's me who picks the team.

Outside, on a busy main road in the rush hour. A single pair of adult jogging feet. Slight breathing and the loud, close honk of a car horn.

TEACHER. That's Bill. Electric-blue Audi. Passes me every morning.
Big joke. Wait till he's down and out with a coronary. Two miles,
it is. Pretty flat. Just the one hill. Started September. Too fat.
Sweating like a pig when I get to school. Tracksuit on all day. Kids
like it. Some kind of hero. Thinking of having a go at the Festival
Marathon next year.

The staffroom.

TEACHER (*calling*). Bill . . . hey, Bill. See our test results? IQ.
Intelligence Quota. What about Terry? Didn't score. Bet some of
your lot did better than that. Should be in your class, Terry should.
Come off it, Bill, he's as thick as eight short planks. Reading age of
one and a half. Only thing he understands is a clip round the ear.
Sixteen? I've got thirty-two. Two over the limit.

Tell you what — you can have Roger Burge as well. He needs help.
I need help not strangling him. Oh no, no swapping. I'm over the
limit already. You take Terry and Roger. As a gift.

Cross-fade to corridor. There are no children.

(*Under his breath:*) They don't like me, see. Face doesn't fit. Won't
join the union. Don't bother me. I'm paid to teach kids, not join a
social club. All so bloody wishy-washy.

Lose corridor acoustic.

Whole food, bran, CND . . . you go into their classrooms. Chaos.
Total shambles. Sixteen, Bill's got, and he can't control them. Then
they wonder why the kids don't respect them. Kids need discipline,
security. What's that book, on a desert island — *Lord of the Flies.*
It's true — like animals if you let them go. Tear you to pieces. One
sign of weakness and they'll have you.

*On the tarmac outside school. Adult pouding feet close to the
microphone and children's feet further off.*

TEACHER (*breathlessly*). Run, Adrian. Work. That's good, Carlton,
but keep it up. Don't slack off. (*He stops running.*) No use sprinting
just when I'm looking. Sprint all the way. (*He spits:*) Look at him.
Roger the Dodger. Not even puffed. Are you puffed, Roger Burge?
Are you? Well you ought to be. If you're not puffed it means you
haven't done your best. You've got to run your hardest every time.
If you don't run flat out, you'll never improve — that means get
better — understand? Flat out means you end up flat out on the
ground after every run — dead? Get it? In the district next week
you'll be up against some real runners, not these spasos. Now get

round again. (*Calling after him.*) And this time *run*.

In the school dining-hall. Not too noisy.

TEACHER (*eating*). Trouble is, you can't blame him. No competition. Gets bored, beating the same old rubbish. He knows he's best, they know he's best. We all know it. Needs someone to push him. Find out what he's got left over. Terrible, isn't it? The food. Free, though — if you do your duty. Saves cooking at home. You know, I've never seen him stretched. Not once. Never seen him really suffer — even beating kids three years older. Seen him go hard, but always fresh as a daisy in the end. In this one race — district it was, last year — let this red-headed kid get a hundred yards on him first lap.

Fade down canteen. Fade up exterior, feet through mud, excited children's voices cheering.

Pulled it back and they were neck and neck coming up the hill at the finish. Thick mud, first Roger with his nose in front, then this other kiddie. Screaming my head off I was.

TEACHER's *voice emerges from background.*

Come on, Roger. Work, run. Crush him. Take him now.

Fade exterior background. Gain a little canteen.

This kid held on, and Roger held on, and I thought his heart was going to break. Face like a mask, screwed up — like Zatopek. Wasn't rolling, though. Running smooth, like an animal. Can't break that kid's style. And then, twenty yards to go, he kicks, and the red-head's beaten, gone. Roger? Hands on his knees a minute blowing, then he's off to get an ice-cream. And when they race again in the county, the red-head's broken. Can't touch Roger. Never will again. That's what Roger's like — a winner. He won't take second best. Sprinting's totally different. There's your best, and you always get within a tenth or so. Used to know every race how it'd finish, unless someone pulled a muscle, or false starts. No point betting against the favourite. That's the thing about sprinting. No tactics, no dark horses, no upsets. Just getting beat. Ten, three, wind assisted. That was my lot.

(*Loud stage-whisper.*) Here, Roger. Pssst. Afters. Crumble and custard. Want it? (*Quieter whisper, since Roger has come over.*) Here. Don't let Mrs Crask see.

Plate noise.

She'd murder me if she knew I gave it him. Make such a fuss about having it, I do.

Pause.

Needs it, don't he? Never gets a square meal at home.

Pause.

Not much I can do for him, really. (*He shouts:*) Kevin Bailey take your own plate back.

Feet plodding along busy main road.

TEACHER (*slightly breathless*). Next year he'll be off up the big school. Comprehensive. Submerged. Two thousand kids. Another year or two, fags and red stripe and hanging about on corners. Drugs. Stones at the police. See them every day, you do, after they've left here. All the sweetness gone out of them. All the energy. Flabby. Just enough strength to steal an old woman's handbag.

No chance — don't touch the stuff. Never have. Never interested me. Or fags. Separated. She took the kids.

Bill's horn.

Sod off, you bugger. Always offering me a lift, he is. Trying to tempt me, so he can have a laugh. Like I say, I'm in no hurry to get home. Done very well for herself. Took me to the cleaners. Had to sell the house. Suppose she didn't like me. Got a flat now. More of a bedsit. Share the kitchen and bathroom. Bachelor sort of place. Could be worse. Neighbour's a pain. Young. You know. Really want a place nearer the school. Wouldn't have to run so far. Can't afford it, though.

An empty slightly echoey classroom. Distant sounds of playground noises.

TEACHER. I'm not pleased with you, Roger. Not pleased at all. You haven't had a very good week, have you? . . . Have you?

No, sir.

Stand up straight when you talk to me. Call yourself an athlete, slouching like that? And get your fingers out of your mouth. You're not a baby, are you? Not in Miss Larkin's class? That's better. Right. You haven't had a very good week. Nor have I. Every time I look round you're up to something. Picking on people. Running round my classroom. Breaking windows. Chatting. Or else you're sulking. In one of your moods. Now listen to me Roger Burge, and I mean listen.
Are you listening?
Are you?
Yes what?
Yes, sir. You think you're special, don't you? Just because you can

run fast and you're tougher than the other louts in the class, you
think you're special.
Don't you?
Don't shake your head at me. Look at me when I'm talking to you.
You think you're special. Well you're not. Not to me you're not.
To me you're just a pain in the backside like the rest of them. I've
got thirty-two little pains in my backside, and I treat them all the
same, like pimples. If they play along with me, then OK. But if they
play me up, then I squeeze them. Get it? I squeeze them hard, until
they burst. Then they're no trouble. It's up to you Roger. If you're
going to make trouble for me, then I'll make trouble for you. Big
trouble. I'll hit you where it hurts and you won't know whether
you're coming or going. I can make life hell for you.

(*He shouts.*) Stand still.

For a start, any more trouble and you'll be out of the team. You
think I'm bluffing, don't you? Think I can't do without you. Well
I can. I don't care if we win that cup or not. I care about the way we
do it. And if you're going to mess about and please yourself, then
I don't want you. Hear me? Do you?

Bell rings loudly.

Yes what?

Yes, sir.

(*In an easier manner, packing some books.*) I don't want clowns in
my squad, see? I want runners. It's a big world out there. You're
the best in this school, but that's nothing. If you want to get
somewhere, beat the big world, you'll have to buck your ideas up.
It's up to you — I can't force you. Understand? (*With humour.*) Yes,
sir. Now get out onto that field and run. Run until I tell you to stop.

No acoustic. He speaks near the microphone, intimately.

Always needed bollocking, that kid. No use mothering him. Got too
much of that at home. Spoilt him rotten, his ma. Funny, her a
whore and all. Do anything for her, he would. When she got done,
ended up at his gran's. Old Ma Cox. Remember her? Right old
battle-axe. Came down and had a go at me with her umbrella once.
Knew how to handle Roger, though. Bawl him out and he'd go. Tell
him he was chicken and couldn't beat so-and-so. Christ, he'd go.
Clenched fists, tears in his eyes: not silly, though. Not rushing it or
anything. Cold, calculating. Racing brain, he had. About the only
brain he had. Roughest of the lot, he was — Carlton, Brian —
remember that gang? Never a leader, though. Never wanted to be a
leader. All the other louts were frightened of him. If he got going,
they'd back down every time. Right little loner, but they had to
keep in with him. Never bore a grudge, though, Roger — straight in
and straight out. Never a bully. Give him half a chance to cool down,

and he was as good as gold. And always stupid little things that got him going.

Cut to locker room. A fight is in progress. The sound of blows and rolling bodies against lockers. An excited audience of children.

TEACHER (*bursting through a swing door*). What on earth's going on here?

The audience silences but the fight continues.

Brian, Roger, cut it out. Roger (*forcibly separating them.*) get off him. Brian, go and see the nurse. The rest of you outside. (*He shouts:*) Now.

Ebbing crowd, whispering.

(*Calm, almost shocked.*) What on earth was all that about?

Pause.

Here, blow your nose. No good crying now.

Pause.

Beating up my centre forward. I need him in one piece for Wednesday. Could have killed him.

Outside, on the main road in rush hour. Feet pounding.

TEACHER. Can't play any favourites in this job. Just 'cos his old woman's a whore and he's the best runner the school's ever had. One rule for everyone. Treat him just like the spasos. Gave him a belting. Water off a duck's back. Shows the others, though.

Patrick — he runs every day. Keeps up with the girls. Has a go. Does his best. What's the difference. Give it a couple of years -- all the power, energy — discos, fights, girls. Drinking, smoking. What's the odds?

Outside at a football match. Some children are chanting 2 — 4 — 6 — 8. Occasional whistles and the sound of boot on ball.

TEACHER (*calling from the touchline*). Tackle him, Jamie, tackle him.

(*Aside:*) Worst football team we've ever had. Can't really call it a team. Only Carlton can play, and he runs backwards every time he gets the ball.

(*He calls.*) Get back, Roger.

(*Aside.*) He's no player. OK when he's in the mood — chasing and tackling. Just needs one thing to go against him, and his head goes down, and that's it.

(*He calls.*) Clear it, Bobby, kick it.

The half-time whistle sounds.

(*He calls.*) Right, gather round. Roger, come here. Carlton, Jamie.

That was diabolical — diabolical, do you hear? I should have picked the girls' team. Carlton, if I see you running backwards once more I'm going to kick your backside from here to Basingstoke.

And Roger — what's your name — Shirley Temple? Wandering about half-asleep. Gone all moody, have you, just because I had a go at you yesterday? Well, you'll just have to get used to it — because I'll be having a go at you every day until you buck your ideas up. Get out there and work. I want you to kick everything that moves. I don't want to see anyone or anything getting past you. Get it?

What was that?

Look, if you don't want to play, don't play. There's plenty of boys who do. Do you want to play or don't you? Well? Right. Go and get changed.

(*He calls.*) Patrick, you're on. Big moment. Go on, Roger, get off. You've had your chance.

The whistle blows.

The rest of you get out there and get stuck in . . .

Outside, on the main road in the rush hour, feet pounding.

TEACHER. Four-nil, in the end. St Joseph's too. Half the size of our lot. Roger sulking, affects the whole team, see. He gives up, the whole lot give up. Should've heard Bill. Comes up to me after the match. 'I've been talking to Roger.'

One up to him for a start. Bugger won't talk to me. On about his background. I told him. We've all got backgrounds. Can leave their backgrounds at home when they come into my class. No time for it. Play at being victimised somewhere else. Going on about how we'll lose the cup if he's not in the cross-country. So what? Privilege to run for the school, not a right. Told him there and then. 'You want to pick the team, train them, ferry them about, wipe their noses, you're welcome.' Course he wasn't interested. Too much like hard work. Had a chat with him. Wouldn't have him in his class, though.

Crossfade to walking along empty corridor, approaching a fairly noisy classroom.

Yes, I'm hard. But you've got to be hard. It's for his own good. He's got to learn. You can't make exceptions. If he behaves himself, he'll be in the team. Knows where he stands.

He enters the classroom. The noise subsides.

Right.

(*Louder.*) Right, Brian Jennings.
Carry on with your maths work books.

A groan.

Quietly. (*He sits down.*) Half of it's he wants attention. But what about the rest? See you treating him different because he's playing you up — they all start playing you up. Why should the trouble-makers get all the attention? Go away, Bobby, I'm busy.

Special help, he needs. Don't we all? In the mood, he'll work. Got to find something he understands, though. Maths — clueless. Look at him — in his trance. Sent them on a matchbox treasure hunt, once. Sunny day, get them out of the classroom. Got to find as many different things as you can that fit into a matchbox. Great. Roger's off like a whippet. Finds a leather-jacket — woodlouse — sticks it in the matchbox. Then one of the girls says it's cruel. So Roger gets it out, puts its eyes out. Very careful, mind. Pinches them. 'If he can't see where he is, he won't mind being shut up in the matchbox.' Crazy. Really fond of that leather-jacket he was. And his team won the treasure hunt.

Crossfade to the sound of running.

Then there was the kite. Got in early one morning; found him out on the field with this old fan belt and a broken stick. Making a kite, can I help? God knows where he found the fan belt. Found him some sugar paper, Sellotape, helped him stick it all together. There he is, wind blowing up a gale across the field, crisp packets, sweet papers flying everywhere; running backwards and forwards trying to get this abortion of a kite up in the air. Didn't know whether to laugh or cry. Left him to it. Never flew, it didn't. Ended up hitting people with it. Got a belting.

Gradual fade of feet and breathing into the distance.

In the classroom. Lassitude and low painting noises.

TEACHER (*casually as he moves about the room*). Very nice, Jennifer. What's that green bit? (*Moving away.*) OK, OK, fair enough.

(*Abrupt, intimidating.*) Yes, Bobby.

(*Explaining slowly, impatiently.*) For the last time, I'll put the team

up on the board after training tonight. OK? So anyone who doesn't
come running tonight won't be in the team tomorrow.

(*Overcoming objections.*) I told you there was training tonight —
told you on Wednesday. You know the score. In this class you
always bring your kit. We've been through all that. Just get on with
your work. You all want lovely paintings to take home to your
mums, don't you? Anyway, Bobby, I don't see why you're so
interested in the cross-country. You couldn't run a bean, let alone
a race.

Giggles.

Paint!

(*Sitting down.*) Friday afternoons. Art. Can't stick it. Don't mind
them drawing, but when the paint comes out — the mess. That's why
I have it Friday. Give the cleaners the weekend to clear up. Usually
tell them a story the last half hour. Gives me something to do.
Before I run home. Look at it — peeing down.

(*He calls.*) Roger Burge! Did I ask you to get up? Did I? Well then,
sit down. That's better. Put your hand up if you want something.
Right. What do you want? Shouldn't need a rubber. We're painting,
not drawing. Let's have a look. Hold it up.

Giggles.

No comment. Anyway, I haven't got a rubber — don't need one.
Never make a mistake, do I?

Groans.

Anyone got a rubber they can lend Roger Burge? By the look of his
picture he needs a big one.

Laughter.

Must be mad, Catherine. Or madly in love.

Titters.

There you are, Roger, and don't ever say we don't look after you.

Pause.

Typical Friday afternoon. Look at him. Sprawled out, face all
scrunched up on his fist. Pulling poor old Catherine's rubber to
pieces. Next thing he'll be flicking it round the room. Then he'll
be sticking his paint brush up Bobby's backside. Give him his due,
he's done his best the last few days. Sullen, but subdued. Have to
have another go at him in a minute. Longer you leave it, the worse
it gets. Trouble is, I can't be bothered. (*Standing up.*) Right. If you
haven't finished your painting, then carry on. If you have finished,

then you can listen to me. Or you can do both.

Roger, if you don't want to listen you can get out. It's up to you.
Thank you.
Tomorrow is the cross-country, and today I'm going to tell you the
story of a very famous runner from history.
He lived in a country called Greece many, many years ago — long
before Jesus was born, even — and his name was Pheidippides.

Chalk on the blackboard.

Pheidippides lived in a part of Greece called Sparta. Roger Burge I
can see you.

Pause.

(*Quieter.*) He was a young boy, about your age. A runner, a
messenger in the Spartan army. The Spartans didn't believe in
writing or sums, they — Roger Burge, I warned you. Come here.

(*He shouts.*) Come here.

Own worse enemy, aren't you? Give me that rubber. Now get out.
Go on. Go and see the headmaster. I haven't got time to waste on
you. (*He shouts.*) Go on, get out of my sight.

The door opens and closes. Pause.

Pheidippides, when he was about your age, he left school and joined
the army Adrian turn round . . .

Fade down.

On the road. Running feet.

TEACHER. Some days you think you've got him — got through to him.
On the same wavelength — do anything for you. Seen him smile?
Lights up his whole face. Smiles like he means it, that kid. Next
minute he's away. Forgets. As if you didn't exist. Not like most kids.
Tell them off and they do what they're told all grudging and huffy.
Not Roger. Coils up inside himself. Talking to a brick wall. Old
Parkie had him in there all afternoon.

(*He imitates the headmaster.*) 'You're not leaving this office until
you tell me what's the matter.' Never said a word. Not a dicky bird.
All afternoon, he was in there. In the end he sends him back to me.

*In the locker rooms: high-spirited sounds of changing of clothes and
shoes.*

TEACHER. Training he starts getting changed as per usual, in the
corner, on his own. What do you think you're doing, Roger Burge?

Oh no you're not. You can get your shoes back on and get off
home. Now. I don't have clowns in my team.

Pause. In the staff room.

TEACHER. Didn't have any choice, did I? Can't make exceptions.
Never came running again. Too proud. That was his trouble. Never
admit he was wrong, say sorry. Hanging around in the playground,
watching the others run. Came nowhere in the district, without him.
Hopeless. All the others reverted to their natural state — spasos.
Wasn't worth entering a team. OK in class though, after that. Good
as gold. Kept himself to himself — occasional outburst, but nothing
much.

A bell sounds.

Never did any work, mind.

(*He calls.*) Got your car today, Bill?
Come off it, it's peeing down.

The door opens to the sound of children's babble.

Meet you in the car park — (*He calls.*) and wait for me this time.

Fade up babble to unbearable level. Cut suddenly.

SCOUTING FOR BOYS

by Martyn Read

Martyn Read was born in Henley-on-Thames in 1944. He had a variety
of jobs (including grape picking in Spain and potboy in an Oxford
college) before becoming an actor. He has worked extensively in
television, theatre and radio, for which he began to write six years
ago. Radio plays include: *Thank You For Your Support*, *Where Were
You the Night They Shot the President?* and *Waving to a Train* which
won the 1980 Giles Cooper Award and which was adapted for BBC
television and shown earlier this year. His one-man play *221B* about
Dr Watson was performed by Nigel Stock at the Edinburgh Festival
(1983) and subsequently toured nationally.

Scouting for Boys was first broadcast on BBC Radio 4 on 27 June 1983. The cast was as follows:

TIGER TIMMS *sixty; vigorous, brisk, deep voice;* *not many teeth perhaps.*	Nigel Stock
COOPER *known as 'Coop'. Mid-twenties, seems* *older. Neutral accent.*	Alex Jennings
GEORGINA) MILES) *twins. Upper class archetypes.* *About thirty.*	Alison Steadman Jeremy Child
THE COLONEL *their father. NOT a caricature army* *type. A rather gentle, liberal man.*	Michael Spice
SPIDER *young scout. About 17/18. Voice* *sounds as though it's only just broken.*	Spencer Banks
YOUNG MILES) PRINGLE)	Elizabeth Lyndsay
YOUNG GEORGINA) TUCKER)	Jill Lidstone
YOUNG SCOUTS	Lloyd Charlton, Jason Cooper, Adam Rhodes, Timothy Ward

Director: David Spenser

The song *Ging Gang Gooly Gooly Watcha* is a Scout campfire nonsense song.

All passages within double quotation marks are verbatim quotes from *Scouting for Boys* by Robert Baden-Powell.

Fade up scout song 'Ging gang gooly gooly watcha'. Hold under first speech.

TIGER (*internal*). "I was a boy once. The best time I had as a boy was when I did a good lot of scouting in the woods in the way of catching rabbits and tracking animals and so on. Well, I enjoyed this kind of life so much I thought 'Why should not boys at home enjoy it too?' I knew that every true red-blooded boy is keen for adventure and open air life, and so I wrote this book to show you how it could be done."

Fade song. Mix to a motor coach stopping and engine cutting out. Door slides open. TIGER gets out onto grass. Full summer sound effects.

Stretch the old legs! Flex the old knee joints! Breath *in*!

He inhales and exhales.

COOP (*calling*). We stopping here, Tiger?

TIGER. Dunno, Coop. Down the lane, wot d'you reckon?

COOP (*approaching*). Looks all right to me.

TIGER. Bit of a wood, an' a stream. Worth a recce.

COOP. I could do with cooling off.

TIGER. 'Ot?

COOP. Hard work driving, a day like today.

TIGER. She's a good ol' bus though, ain't she?

COOP. Demanding.

TIGER. Bin all over with us. We was lucky to find 'er when we did.

COOP. We were even luckier to find a patrol leader who's a bit of a wizard with seized up Bedford engines.

TIGER. Yeh all right, thanks very much. Got us up the old A1 though, didn't she?

COOP (*with a little laugh*). The old A1.

TIGER. You can laugh. The Great North road. Got a ring to it boy.

SPIDER (*from the coach*). Tiger!

TIGER. What is it, Spider?

SPIDER. Can we get out? Only it's jolly baking hot in here!

TIGER. 'Adn't forgotten you, lad. Jump out. You can all come and 'ave a paddle.

SPIDER. Hear that, Scouts? A paddle!

ALL (*from the coach*). Hurray!

TIGER. Bring the mugs and a billy. We'll have a brew up.

SPIDER. OK, Skipper.

COOP. Mmm, feel that sun.

TIGER. You know, I ain't 'alf been lookin' forward to this, Coop. Weekend camp. Never palls. Smell that air. Drink it in lad. "Drink it into your lungs and blood."

They breathe in deeply.

COOP. What about this recce?

TIGER. Right. Race you to the stream!

COOP. You're on!

Fade accoustic down but not out.

TIGER (*internal*). "Camping is the joyous part of a Scout's life. Living out in God's open air among the hills and trees and the birds and beasts — that is, living with Nature, having your own little canvas home, doing your own cooking and exploration — all this brings health and happiness such as you never get among the bricks and smoke of the town."

Fade up sounds of splashing and laughing.

COOP. Stop splashing, you old devil! I'll shove you under!

TIGER. You an' whose navy, eh?

He splashes COOP.

COOP. Right, you asked for it!

Laughter and splashing.

TIGER (*internal*). "Camp Fire Yarn Number One. When you go camping you must first decide where you will have your camp. To my mind the best place is in or close by a wood where you have permission to cut firewood and to build huts. Remember that a good water supply is of first importance. If there is a spring or stream, the best part of it must be kept strictly clear and clean for drinking water. Farther downstream a place may be appointed for bathing, washing clothes and so on."

COOP. Pack it up!

TIGER. Lovely though, innit? Cool water runnin' through yer toes.

COOP. Tastes good too.

TIGER (*slurps*). Ambrosia.

COOP. Nectar.

TIGER. Nectar then, clever dick. This'd be all right for the ablutions here, wouldn't it? Further up for the drinkin', and down there, we could 'ave that for the nood bathin'. Teach the new ones to swim.

COOP. It's an ideal site. You certainly know how to find them.

TIGER. Timm's principle. Never fails: 'If you keep turning to the left, you're bound to end up somewhere interestin'.

COOP (*laughs*). You said it!

TIGER. Oh — on to the old dooble ontondries already are we? The old interlectual stuff!

COOP. Blame it on night school!

TIGER. Turnin' to the left. You're too sharp for me.

COOP. That'll be the day!

They laugh. Fade so we hear them in the distance.

MILES (*close*). Well, well, what have we here? A brace of Boy Scouts! Come to paddle in my stream and play in my wood. Got them in my gunsight. One young fledgling, nice and juicy, and one tough old bird. I wonder if it's closed season for Boy Scouts? (*He giggles.*) Lie low, Miles, watch them, stalk them. (*He giggles again.*)

TIGER. Wonder who this place belongs to?

COOP. There's a notice, look, on the far bank.

TIGER. You're sharp. I never saw it.

They wade across.

COOP. Not surprised with those glasses.

TIGER. Nothin' wrong o' National 'Ealth specs, mate. Fifteen bob thirty years ago. See me out.

COOP (*whistles*). Listen to this.

TIGER. Woss it say?

COOP. 'Private Property. Trespassers will be hanged, drawn and quartered.'

TIGER (*sharply*). What's that you say, boy?

COOP. That's a bit strong, isn't it?

TIGER (*quietly*). 'Anged, drawn and quartered . . .

COOP. You all right, Tiger?

TIGER. Gyppos can't read.

COOP. Tiger?

Cut.
The sound of a wooden stake being hit into the ground.

COLONEL. Hold it steady man!

TIGER (*twenty years younger, but sounding much the same except more belligerent*). I ain't sacrificing my fingers to you an' all!

COLONEL. As well as what passes for your soul, you mean? (*Final hit.*) There.

TIGER. 'Trespassers will be 'anged drawn and quartered'. Bit strong.

COLONEL. Gets the message across.

TIGER. Gyppos can't read, any'ow.

COLONEL. Did you lock the church? I don't want them pilfering.

TIGER. Yeah. I don't know why you don't just sling 'em off.

COLONEL. I don't want them over here where the twins can associate with their children.

TIGER. Lowers the tone, don't they, tinkers?

COLONEL. But anyone is perfectly free to camp the other side. That always has been, and always will be, common land.

Cut back to the present and the sounds of the stream.

COOP. Tiger, hey!

TIGER. Mm? What?

COOP. I said we'd better go back.

TIGER. Sorry, Coop. Day dreamin'.

COOP (*with real concern*). You OK?

TIGER. Yeah. The 'eat, I expect. On me 'ead. Wivout me 'at.

COOP. Let's get you on the bank.

TIGER. Be all right when I've 'ad half a gallon of ol' Spider's tea.

They climb out. MILES watches.

MILES (*quietly, with barely contained excitement*). I spy with my little
eye something beginning with T! Oh golly, what have you flushed
out from its cover, Miles? A rare old bird! Whee! Oh gosh, late for
dinner. Georgie'll be cross. Better get a couple of rabbits. But wait
'til I tell her I've been after bigger game. Wait 'til I tell her I've
bagged a Tiger!

Cross to TIGER and COOP on bank.

COOP. Got your boots?

TIGER. Good old boots.

The sound of a pheasant alarmed.

COOP. Hello, someone there?

TIGER. Creature of the woodland, that's all. You ready?

COOP. Just roll my trouser legs down.

TIGER. I wish you'd wear shorts, you know.

COOP. I'm too old.

TIGER. I wears 'em. "Shorts are essential to hard work, to hiking and
camping. They give freedom and ventilation to the legs."

COOP. And make me feel a twerp.

TIGER. No Scout need ever feel that.

A shot is heard and birds alarmed.

TIGER. 'Ullo.

COOP. Thought there was someone about.

TIGER. Out for a bit of mornin' sport that's all.

COOP. Poor old creatures of the woodland.

A second shot is heard.

Ah, the jolly old double-barrelled twelve bore, what?

TIGER (*quietly*). Holland and Holland. Side-lock ejector.

COOP. What?

TIGER. Takes you by surprise.

Cut dead to close interior accoustic.

COLONEL. And this most importantly, Miles, is the safety catch. (*He clicks it.*) On, off.

TIGER. Ain't right, teachin' the boy bloodthirsty 'abits.

COLONEL. It's a fine gun. Holland and Holland. Be yours when I'm dead and gone.

TIGER. Roll on.

COLONEL. Think you can manage it?

YOUNG MILES. Is it safe?

TIGER. 'Course. Give it 'ere. I'll show you.

COLONEL. Steady, Timms, the catch —

TIGER. You can squeeze as 'ard as you like. See?

COLONEL. You idiot!

YOUNG MILES. Daddy, Daddy, Daddy!

TIGER. Oh dearie me.

COLONEL. You've blown a damned great hole in the floor!

TIGER. Yeah. Sorry about that.

COLONEL. It could have been my foot!

TIGER. The one that ain't already in the grave, you mean?

Cut back to outdoor acoustic.

COOP. What's up? You dreaming again?

TIGER. Coop, this place —

COOP. What about it?

TIGER. Oh, nothin'. I like it 'ere already. Feel at 'ome.

Distantly comes the sound of a human trying to imitate a Curlew.

COOP. Listen, there's old Spider. Tea's up.

TIGER. 'Is call don't get any better do 'it? 'Ang on. (*He replies. If anything, worse.*) Now that, you see, is impossible to distinguish from the real thing. (*They laugh.*) I enjoyed me paddle, Coop.

COOP. Me too. Spider's coming on, isn't he?

TIGER. All your doing. 'E looks up to you.

COOP. "The Cub looks up the the Boy Scout, and the Boy Scout looks up to the Old Scout."

TIGER (*embarrassed*). Yeah, well. You're a good patrol leader, Coop. I'm proud of you all, you know. You, Spider, the new lads. The movement's growing all the time.

Cut acoustic.

TIGER (*internal*). "The object of becoming an able and efficient Boy Scout is not merely to give you fun and adventure. A true Scout is looked up to by other boys and grown-ups as a fellow who can be trusted, a fellow who will not fail to do his duty however risky and dangerous it may be and a fellow who is jolly and cheery no matter how great the difficulty before him."

Cut to the camp clearing. A small fire burns. Laughter.

TIGER. Good cup o' tea that, Spider. Very nice.

COOP. I don't know how you can drink it as strong as that.

TIGER. Ganges tea. Wiv a slice of bread an' marge, delicious!

SPIDER. Sweet enough?

COOP. He's got half of Tate and Lyle in there!

TIGER. Man's drink, mate. None o' this dishwater you sip at.

COOP. I've got highly developed taste buds.

TIGER. Oh ah. You'd've 'ad a rough time in Ceylon, then.

COOP (*laughs*). Here we go.

TUCKER. You been abroad, sir?

COOP. Don't ask, Tucker.

TIGER. Course I 'as. Stoker First Class in the Merchant Marine. The old Red Duster. We used to call in at Colombo, take a load of tea on. Boil some up. Fresh as a daisy it was. Deep, dark orange colour. In a tin mug, lovely.

TUCKER. Ugh.

TIGER. Course, ol' Coop 'ere, 'e's heducated, been to college. 'E likes all this scented muck with your little finger crooked. (*He laughs, a sort of cackle. Others laugh with him.*) Mind you, I bin in places like that. (COOP *and* SPIDER *groan.*) It's true.

PRINGLE. Tell us, sir.

TIGER. I bin in places where . . . where they 'ad this stuff 'Earl Grey'. Know what it was? Gnat's widdle. (*Others giggle.*) 'Jest top up the pot, Timms,' they used to say. Blimey, weak? And sandwiches — weren't worthy of the name. Little triangles wiv no crusts with this stuff 'Gentleman's Relish'. Tasted like boot blackin'. White linen, all starched. China that thin you could see yer 'and through it. Fancy biscuits. (*Small pause.*) Bit o' mint cake.

PRINGLE. Where was that, sir?

TIGER. Mmm? Oh that was . . . a long time ago, Pringle lad. (*He slurps tea.*) That's a good lookin' fire there. Who done it?

SPIDER. Tucker, Pringle and some others. I supervised.

TIGER *and* COOP *laugh.*

TIGER. Got to be careful of fire though. Everythin's tinder dry this time of year.

SPIDER. We stoppin' here, Tiger?

TIGER. 'Ope so. Got to see about permission first.

COOP. D'you hear those shots, Spider?

SPIDER. Yes. I think they came from over by the church.

TIGER (*sharply*). Church?

SPIDER. Up the stream, other side of the wood there's a little church.

TIGER. What's it look like?

SPIDER. All overgrown.

TIGER. D'you go in?

SPIDER. We were too busy collecting wood.

TIGER. Should 'a bin locked. We don't want 'em pilferin'.

SPIDER. Pardon?

TIGER. No one about? No vicars, curates, things like that?

SPIDER. It was deserted. Can we have our paddle now?

TIGER. All right, lad. Cut along.

SPIDER. Come on Tucker, Pringle, I'll teach you the patrol call.

TIGER (*chuckles*). Old Spider. Reckon 'e knows the book better 'n me!

COOP. Never!

TIGER. Come on then. Sooner we get permission, sooner we can set up camp.

COOP. Do we have to?

TIGER. What?

COOP. Well, it's so secluded no one would ever know we were here.

TIGER. I'm surprised at you, Cooper. You know what the book says.

COOP. "Be careful to get permission from the owners of the land before you go on it." But they never object. And suppose we can't find them?

TIGER. Then we moves on.

COOP. And pass up a perfectly good campsite.

TIGER. Don't argue with me, Coop. You — you know what it does.

COOP (*beat*). Yeah. All right, let's see if we can run our sporting friend to earth.

Cut acoustic.

TIGER (*internal*). "Camp Fire Yarn Number Two. The importance of tracking. Tracking or following up track is one of the principal ways by which Scouts gain information and hunters find their game. You can learn about different animals by following their tracks and creeping up to them so that you can watch them in their natural state and study their habits."

Cut to stately home dining-room. Slight echo. Dog snuffling, then barking.

GEORGINA. Enjoy that Rufus? The young master's lunch? Serve him right, he knows the rules. (*A clock strikes.*) Half an hour late. Where *is* my impossible brother? He *knows* the trouble I go to. Finished? Think you could manage some pudding?

MILES (*off*). Georgie!

GEORGINA. Ah, the return of the prodigal.

MILES. Georgie, Georgie!

GEORGINA. Wipe your shoes before you come in here!

MILES. Botheration. There. (*He approaches.*) Georgie, I've something ripping to tell you!

GEORGINA. Where on earth have you been?

MILES. Red hot news! Absolutely whizzo!

GEORGINA. You're not telling me anything until you've calmed down and explained your lateness in coming to table.

MILES. I've got a jolly good excuse!

GEORGINA. There is no excuse for bad manners. You know I do not like to eat alone. Have you washed your hands?

MILES (*exasperated*). Oh, yes.

GEORGINA. Show.

MILES. There. I washed them in the stream.

GEORGINA. Oh, we've been down there have we? Playing Cowboys and Indians?

MILES. I was out shooting *actually*. Got a couple of rabbits for the pot tonight. Here.

GEORGINA. Not on the table, Miles! Oh God.

MILES. Nice and plump.

GEORGINA. I know we are a once noble family living in reduced circumstances, but I object to my polished rosewood looking like an eighteenth-century still life.

MILES. Oh, never mind the rabbits, let me *tell* you!

GEORGINA (*patiently*). Miles, I know you are a little deaf and I hope I make ample and patient allowance for that — *But will you bloody well sit down and eat!*

Pause.

MILES. All right, if that's how you feel.

GEORGINA. It is!

MILES. Shan't tell you. See if I care.

GEORGINA. Shut up. Here.

She passes a plate.

MILES. What's this?

GEORGINA. What does it look like?

MILES. Boiled potatoes and carrots? Where's my mince?

GEORGINA. Your lunch, dear brother, is in the dog.

MILES. You beast! Twelve o'clock's a silly time for dinner, anyway.

GEORGINA. If twelve o'clock was good enough for Mummy and Daddy it's good enough for us. Do you want any pudding?

MILES. What is it?

GEORGINA. Semolina.

MILES. You said it would be roly-poly!

GEORGINA. You're lucky I can produce anything from that wreck of a kitchen.

MILES. Honestly, Georgina, it's not much to ask. You look after the house and I run the estate.

GEORGINA. Run the estate, you! Squire Alltweed, benevolent seigneur, patrolling a few acres of decrepit parkland? . . . There.

MILES. It's got lumps in it.

GEORGINA. Give them to Rufus.

MILES. Mangy cur!

He kicks. The dog yelps.

GEORGINA. Honestly Miles, I wonder what an anthropologist would make of you? Slurping semolina with a tweed cap on, and your elbow in a rabbit. A doomed species in its natural habitat.

MILES. Eh? What? Who's doomed?

GEORGINA. God, you're disgusting. I don't know why I put up with you.

MILES (*beat*). Oh yes you do.

GEORGINA. Dear brother.

MILES (*final slurp*). Finished. Can I get down?

GEORGINA. Not yet. I'm waiting.

MILES. What for?

GEORGINA. You had something to tell me.

MILES. Shan't. You spoiled it.

GEORGINA. Red hot, you said. Absolutely whizzo.

MILES. I'm off to skin these rabbits. D'you want the pelts?

GEORGINA. Miles, we know what happens to little boys who sulk, don't we?

MILES. Don't care.

GEORGINA. First of all they get their hair pulled.

MILES. No, Georgie, please. Ow!

GEORGINA. Tell me.

MILES. My ear! Mummy said not to touch my ear!

GEORGINA. And then they get tickled.

> MILES *shrieks. A crash as both tumble to the floor laughing. The dog barks.*

MILES. Get off me, Georgie. *Pax, pax!*

GEORGINA. Tell me Milo, tell little Georgie.

MILES. All right, I give in. (*Laughter subsides.*) Well you see, I was pottering about in the clearing, looking for rabbits, when this bus arrived at the top of the lane and out tumbled — guess what!

GEORGINA. I can't imagine.

MILES. A covey of Boy Scouts!

GEORGINA (*amused*). What?

MILES. Cross my heart. There were about a dozen of them with a patrol leader and a second, and Georgie, a Scoutmaster!

GEORGINA (*beat. Already guessing*). Miles?

MILES. Yes! Well, they went for a paddle and I got quite close and — oh gosh!

GEORGINA. Come on!

MILES. Oh Georgie, the Scoutmaster! Short chap. Cropped grey hair. National Health specs. Baggy shorts. Voice down 'ere!

GEORGINA. Tiger! Are you sure?

MILES. Absolutely pos!

GEORGINA. Tiger Timms back here! Did he know where he was?

MILES. Don't think so. He seemed a bit bewildered.

GEORGINA. He would be. I wonder if he'll come to see us?

MILES. Oh he will. You see I did a clever thing. I laid a trail.

GEORGINA. A woodland trail?

MILES. Yes. He's bound to follow it!

GEORGINA. Good boy, Miles. Come on, let's go to the nursery.

MILES. Oh I don't want a rest. I'm too excited.

GEORGINA. Not a rest. With Daddy's old binoculars, darling, we can watch the Tiger approach.

MILES. And then?

GEORGINA. And then we'll pull his tail!

Cut to the interior of a small church with a stone floor. There is an echo.

COOP. Smells musty in here. Can't have been used for years.

TIGER. More a private chapel than a church.

COOP. Private?

TIGER. All them nobby families had a private chapel and their own vicar.

COOP. Personal insurance for salvation eh? It's a bit of a wreck now.

TIGER (*wryly*). Symbol of the decayin' aristocracy I suppose?

COOP. Funny sort of seat here. It's got a little door.

TIGER. Family pew, that is.

COOP. Look at this. 'In Memoriam. H. St. John Bulstrode, rector of this church, died aged ninety-two years.' Think I'd make a vicar, Tiger?

TIGER. You ain't bald. Vicars 'as to be bald. Now come down out of that pulpit.

COOP. 'Dearly beloved, we are gathered together today —'

TIGER (*with rising anxiety*). Cut that out. This is the 'ouse of God!

COOP. 'The lesson today is taken from —'

Cut to the church full of people, twenty years ago.

COLONEL. — From St Luke Chapter 10, beginning at the second verse . . .

He reads throughout the scene except where shown.

YOUNG GEORGINA (*whispering*). Give me a go with your peashooter, Miles.

YOUNG MILES (*whispering*). Shan't.

YOUNG GEORGINA. I'll pull your hair.

TIGER (*in an unsubtle whisper*). That's enough chatter, Miss Georgina. This is the 'ouse of God.

COLONEL. Timms, do be quiet. I'm trying to read.

TIGER. Beg pardon, I'm sure.

YOUNG GEORGINA. Milo, I bet you can't hit Baldy Bulstrode from here.

YOUNG MILES. What'll you give me?

YOUNG GEORGINA. I'll show you my knickers when we go to bed.

YOUNG MILES. All right. (*He blows the peashooter. A distant grunt is followed by giggles.*) Got him!

TIGER. I warned you. (*He clumps Miles heavily.*)

YOUNG MILES. Ow! My ear!

YOUNG GEORGINA. You beast! You beast!

COLONEL. Oh, do be a brave little soldier, Miles.

YOUNG MILES. Daddy, he hit me! (*He continues to whimper.*)

TIGER. I 'ad to chastise 'im. 'E got the vicar.

YOUNG GEORGINA. You can't hit us. *You're* only the servant.

COLONEL (*reading*). 'And the labourer is worthy of his hire!'

Cut to the nursery. A musical box plays 'Girls and boys come out to play'.

GEORGINA. Miles, do turn that irritating bloody thing off.

MILES. See anything yet?

　Cut music box.

GEORGINA. Not a sign. Where did you lay your trail?

MILES. From the church.

GEORGINA. Good thinking.

MILES. Hear that, Foxhunter? A compliment. Things are looking up.

GEORGINA. Leave my rocking horse alone.

MILES. You're looking peaky old chap. Is she feeding you? I think we should retire him after this season. Put him out to stud.

GEORGINA. God, you're a prick.

MILES. You should know.

GEORGINA. I suppose you really did see him.

MILES. I wasn't drunk if that's what you mean.

GEORGINA. Well, it's a hot day. Could have been a mirage.

MILES. Ha bloody ha. Anyway you've locked the whisky in the tantalus.

GEORGINA. Prudent housekeeping.

MILES. I notice you didn't lock the gin up.

GEORGINA. Miles, in my doll's pram over there you will find a dummy. If you shove it in your gob, you will not only satisfy your arrested development but also my desire for a bit of tranquility.

MILES. Pisspots. Good vantage point, this window. I once caught Tiger with my catapult from here. He was bending over in the rose garden.

GEORGINA (*in triumph*). Aha! Got them!

MILES. Can you see them? I told you, didn't I? I told you!

GEORGINA. The Tiger! I'm sorry I doubted your word, darling.

MILES. I wish I had my catapult now. Wheee!

GEORGINA. Poor old Tiger. He obviously hasn't a clue where he is.

MILES. And the other one?

GEORGINA. A *very* interesting young man.

MILES. Huh. I say, shall we give them tea?

GEORGINA. Yes. We shall receive them downstairs. You know, Capability Brown was such a clever man.

MILES. Eh? What? Who?

GEORGINA. You may recall he kindly designed our park for us. Every turning should present a surprise, he thought. And any minute now, poor old Tiger is going to get the surprise of his life.

Cut to outdoor acoustic. TIGER *and* COOP *on the grass.*

TIGER. Cor! Just look at that!

TIGER. It's the big 'ouse. I knew there'd be a big 'ouse!

COOP. Look, here's another arrow. Broken twigs.

TIGER. You don't need no arrer, boy, look at it!

COOP. You're not going to have another turn are you.?

TIGER. Sorry about that. It was just the smell in that church. Made me . . . dizzy. What a picture, eh?

COOP. The stately homes of England.

TIGER. That'll be where our sportin' Johnnie's gone.

COOP. Here's the end of his trail. Circle of stones with one in the middle.

TIGER. "I have gone home."

COOP. All that, one bloke's home.

TIGER. What's wrong with that?

COOP. Showing signs of wear. That part to the left, just a shell. Must've been a fire.

TIGER. Got to be careful of fire.

COOP. An imposing pile, none the less.

TIGER (*thoughtfully*). Yeah . . . uhh . . .

COOP. Tiger?

Cut to flashback. The COLONEL *and* TIGER *on the gravel drive.*

COLONEL. Well, there you are Timms. Imposing pile is it not?

TIGER (*grunts*). Sure you ain't too cramped?

COLONEL. I get by.

TIGER. So that's the little 'ome you bin on about.

COLONEL. It could be your home as well Timms.

TIGER. Eh?

COLONEL. A thoughtful army has fortuitously retired me at the same time as it sacked my drunken batman.

TIGER. Put up job that!

COLONEL. And now that I'm back here I shall need a butler.

TIGER. Butler! Me? A lick-spittle?!

COLONEL. I doubt if anyone else will take you on.

TIGER (*beat*). 'Ow much?

COLONEL. Fiver a week, all found.

TIGER. Butler, eh? Can I 'ave a green baize door and a bottle of port, like in the films?

COLONEL. If you insist. And I'll provide the uniform.

TIGER. Bleedin' penguin eh?

COLONEL. Take it or leave it.

TIGER (*ungracious*). Yeah. All right.

Fade out. Fade in the sound of feet on gravel.

COOP. Well, are you going to knock or am I?

TIGER. Knock? This ain't a semi-detached in 'Endon. Bells they 'as. That long 'andle.

COOP. A bit antiquated.

TIGER. Coop, where did we go last year? Camping?

COOP. Last year? We went to the New Forest.

TIGER. That's right. Them ponies. And before that?

COOP. The seaside. Westward Ho! in Devon.

TIGER. Yeah. 'Ad that good cup of tea there, didn't we? In that caff. And afore that?

COOP. Is there a prize for three in a row?

TIGER. What? (*He chuckles.*) Sorry, lad. Don't mind me. Well go on. Yank away.

The bell is pulled. Distant ringing is heard and Rufus barks feebly.

TIGER. Someone's in. And again.

Cut to a different dog barking ferociously. The interior of the hall. A louder bell.

COLONEL. For God's sake answer the door, Timms!

TIGER. You do it. I'm trying to catch the bleedin' 'ound!

COLONEL. We can't keep the vicar hanging around at his age.

TIGER. I'm tryin' to prevent old Baldy 'avin' a bloody great 'ole torn out of his clerical backside!

COLONEL. And show some respect for the cloth.

TIGER. After 'e slung that hymn book at me?

COLONEL. Well, you were distributing Marxist literature during his sermon.

TIGER (*catches the dog*). Got you, you bugger.

COLONEL. We'll take tea in the drawing room at fifteen-thirty hours. And don't spill any.

The bell clangs.

TIGER. All right, all right. I'm comin'.

The big front door opens.

TIGER. Wotcher Baldy. We was expectin' you. You've met Hereward, 'aven't you? Hereward, say 'ello to the vicar. (*The dog snarls.*) Hereward: Come 'ere. (*Huge roar then silence.*) Oh dear.

Cut back to the present. The door creaks open.

TIGER. 'Ello. Pardon me. Anyone at 'ome?

RUFUS *barks feebly and pants.*

COOP. Hello old fella.

TIGER. He's nothing like Hereward.

COOP. Who's Hereward?

TIGER. 'Ello. Anyone about?

COOP. Look at the size of this hall. And that staircase!

TIGER. Must be someone here.

COOP. Try that door. Where the dog's gone.

Footsteps on marble are followed by a tap on a door.

TIGER. Er — 'ello?

The door swings open.

Oh! Er — good afternoon, madam, sir. Beggin' your pardon for the intrusion. We rang like, but . . . no one . . . er . . .

GEORGINA (*brief pause*). Hello Tiger.

MILES. Hello Tiger.

TIGER. Hello . . . I . . . er . . . that is . . .

GEORGINA. It's Miss Georgina, Tiger.

MILES. And Master Miles.

TIGER. Miss . . .? I don't er . . . I . . .

GEORGINA. Welcome home, Tiger.

MILES. How *are* you?

Cut back to the hall, a slight echo.

COLONEL. Now children, I want you to meet Mr Timms. He's going to be our butler.

TIGER. 'Ello.

YOUNG GEORGINA *blows a raspberry.*

YOUNG MILES. I've got new boots. (*He kicks.*)

TIGER. Yeow!

COLONEL. Look, my darlings. You'll be spending a lot of time with Mr Timms, so you must try to get along.

TIGER. A lot of time? With them?!

COLONEL. Apart from their lessons with Mr Bulstrode I want you to teach Georgina and Miles what the army taught us: a degree of self-sufficiency and a sense of civic duty.

TIGER. Me? You're kiddin'.

YOUNG MILES. What's civic duty?

TIGER. Well now, that all depends on your point of view.

COLONEL. It's all here in this little book.

TIGER (*reads haltingly*). *Scouting for Boys* by Robert Baden-Powell. Funny 'ats and dyb dyb dyb? Catch me!

COLONEL. Any man who follows it won't go far wrong. Now I'll leave you to get acquainted. And steady with the language, Timms. Goodbye, my darlings, until bedtime.

GEORGINA.
MILES.　　Goodbye, Daddy.

TIGER. *Scoutin' for Boys.* I always knew he was barmy.

YOUNG GEORGINA. My daddy's not barmy.

YOUNG MILES. You're barmy, Timms.

TIGER. Look, I ain't no servant. Everyone calls me Tiger. You an' me's got to get along if I'm goin' to teach you about nature an' that.

YOUNG GEORGINA. We know quite a bit about nature already, don't we, Miles?

They giggle. TIGER *grabs them.*

TIGER. Right. Come 'ere.

YOUNG GEORGINA. Ow! Let go!

TIGER. Now, you're gonna say 'ello nice and polite, ent you?

YOUNG MILES. You're hurting!

TIGER. Say 'Hello Tiger'.

YOUNG GEORGINA (*with effort*). Hello, Tiger.

YOUNG MILES (*with effort*). Hello, Tiger.

TIGER. 'Ello, Miss Georgina. 'ello Master Miles. 'Ow *are* you?

Cut to the present. TIGER *is moaning slightly and* COOP *is patting his face.*

COOP. Tiger, come on now.

MILES. Would he like a glass of water?

GEORGINA. Or perhaps something stronger?

COOP. No thank you. He doesn't touch it.

GEORGINA. Well, there's a turn up.

MILES. What a prang, eh? Just toppled over like a tree! Whee!

GEORGINA. Does he often faint so spectacularly?

COOP. He hasn't fainted. Sometimes he has a sort of fit.

GEORGINA. Oh, he's added fits to his repertoire, has he?

TIGER (*grunts*). Coop . . . Coop. That you boy?

COOP. OK Tiger, take it easy.

TIGER. Coop . . . I thought I was . . . Then this buzzin' and then it all went black.

COOP. Well, they said it would happen now and then.

TIGER. Yeah, they did, didn't they?

GEORGINA. Awfully sorry, Tiger.

MILES. Only wanted to give you a surprise.

TIGER. Miss Georgina, Master Miles. It *is* you.

MILES. Large as life.

TIGER. I can't believe it. Y-You're all grown up like.

GEORGINA. It is usual.

MILES. Found the trail then, Tiger? Hadn't forgotten had I?

TIGER. It was you out shootin'?

MILES. Rather! The old Holland and Holland.

TIGER. Side-lock ejector! I said, didn't I Coop?

COOP. Yes.

MILES. Hullo. Don't think we've met, have we?

TIGER. Beg pardon, allow me. This is my mate, my old china, Patrol Leader Cooper. He looks arter me. And Coop, I'm proud to introduce two old friends of mine. Miss Georgina —

GEORGINA. Good afternoon, Patrol Leader.

TIGER. And Master Miles.

MILES. Hullo, old chap. D'you play poker?

COOP. Er − no.

MILES. Pity. (*Beat.*) Well, here we all are then.

TIGER. I can't get over it. Trust the old A1. You know, soon as I got 'ere I liked it. But the church, the stream, I couldn't sort of focus. And when I said about servin' tea an' that, Well, they all just laughed. Ol' Tiger and 'is stories.

MILES. Talking of tea, I expect you could do with a cup. I'll fetch the trolley.

TIGER. The old trolley. Well I'm blowed!

GEORGINA. Remember the tea trolley?

TIGER. I pushed that a few times, Coop.

COOP. Sorry?

GEORGINA. Tiger used to work for us, didn't you know? He was the butler here.

MILES. And a jolly good butler too.

TIGER. Yeah, I wasn't bad, was I?

GEORGINA. If a trifle original.

COOP. I see.

GEORGINA. And what capricious fate, I wonder, brought you to us?

TIGER. Wasn't planned.

GEORGINA. Ah, serendipity.

TIGER. No, I just stuck me finger in the map.

GEORGINA. Exactly.

COOP. It's a knack for making delightful discoveries by accident.

TIGER. Is it? Can't pull the wool over ol' Coop's eyes. He's bin to night school.

GEORGINA. Really? How interesting.

MILES (*wheeling the trolley*). Tea up!

GEORGINA. And now that you've made the delightful discovery, what can we do for you?

TIGER. Well, why we come, was to ask permission, like −

MILES. Have to speak up, Tiger. Bit deaf in one ear.

GEORGINA. A blow received in childhood, you understand.

TIGER. — er . . . permission to camp in the clearing by the stream, we thought. If you could see your way —

GEORGINA. Oh I think so. No objection, eh Miles?

MILES (*puzzled*). The clearing? Why should we? Stay as long as you like.

TIGER. Well, that really is most generous, ain't it Coop?

COOP. Yes. Thanks.

MILES. Right, everyone for a drop of the old Oolong?

GEORGINA. Forgive any spillage, Tiger. He hasn't quite got your touch.

MILES. Have to do it all ourselves now.

GEORGINA. It's so difficult getting staff these days, Mr Cooper.

COOP. It must be a problem.

MILES. There you are Tiger, soon perk you up. So you've got your own little troop now?

TIGER. A few of us. We 'as a good time, don't we Coop?

MILES. Lots of adventure. God's open air?

GEORGINA. Self-sufficiency and a sense of civic duty? Is that your brief, Mr Cooper?

COOP. I'm sorry?

TIGER. You've got a long memory, Miss Georgina.

GEORGINA. Yes, Tiger, I have.

Slight pause.

MILES. Tea, Miles? Yes, Miles, thank you Miles.

TIGER. You're spillin' it all, lad. Let me give you a hand.

GEORGINA. Might that not indeed make things a little worse?

COOP. Perhaps you'd better not, Tiger.

TIGER. Me brainbox 'as recovered now. I could do the cakes an' that.

MILES. But you're our guest, Tiger.

TIGER. Don't feel right, you servin' me. Go on, I'd like to.

MILES. Well . . .

TIGER. You sit down. Tell you what, I could pretend to knock.

GEORGINA. Now there's a novelty.

TIGER. It'll be just like old times, won't it?

COOP. Tiger —

TIGER. 'Ere we go.

TIGER knocks on the table.

GEORGINA. Come in.

TIGER (*posh voice*). Fifteen-thirty hours, madam. Tea his served.

Cut to TIGER wheeling the trolley. He whistles, shrilly.

TIGER. Tea up, Cocky!

COLONEL. D'you usually barge in without knocking?

TIGER. Usually, yes.

COLONEL. I'm having a private conversation with Mr Bulstrode.

TIGER. He could probably do with a cuppa after what Hereward done to 'im. Here's your marge.

A clunk as half a pound of marge is dropped on the table.

COLONEL. You'll be the death of me. Get out!

TIGER. Suit yerself. 'Ere's yer trolley.

He pushes it at the COLONEL.

COLONEL. That's the best china, man!

Crash and grunt.

Oh my God, now you've scalded the poor old bugger. I'm most dreadfully sorry, Mr Bulstrode. It's so difficult finding staff these days.

Cut to the present. TIGER is serving tea genteely.

TIGER. May hai refresh your cup, sir?

MILES. Rather. Oh, and a cup for yourself.

TIGER. Don't mind if hai do, sir. Lovely tea. Very refreshing.

MILES. Sandwich?

TIGER. If hai may sir, I think hai'll just have a fancy cake. (*In his normal voice:*) I like them ones with the little bit o' pink icin' on top.

General laughter which does not include COOP.

Well, 'ow did I do?

MILES. Jolly good Tiger, bravo.

GEORGINA. Once a butler, always a butler, I say.

COOP. I think we should be going soon, Tiger —

TIGER (*cutting in*). Wot 'appened to Baldy Bulstrode?

MILES. Baldy? Poor old chap. Began to gibber.

GEORGINA. Kept seeing the devil dressed as a butler.

TIGER. Who's the vicar there now then?

GEORGINA. No one. Baldy was the last. After him, they closed the church. No one came much anymore. Apart from the family.

TIGER. The family. I was wonderin' —

GEORGINA. Yes?

TIGER. Well — er — apart from you and Master Miles, well, is there anyone else livin' 'ere like?

GEORGINA. No, just the two of us, rattling around.

MILES. And Rufus here. Mind you, we closed off the West Wing.

GEORGINA. Cheaper than repairing it — after the fire.

COOP. There was a fire then?

GEORGINA. Spectacular fire, Mr Cooper. Flames sixty feet into the sky.

MILES. All old timber. Went up like a bomb.

TIGER. W–what I mean is, are you in charge 'ere now? Is it yours like?

GEORGINA. Oh, I see what you're getting at. Of course, you don't know. He means Mummy and Daddy, Milo.

MILES. Oh, *that*.

TIGER. Well, I was just wonderin' like —

GEORGINA (*matter of fact*). Daddy hanged himself from the main staircase one night.

   Pause.

TIGER. 'Anged 'isself, the colonel?

MILES. With his MCC tie.

   Small pause.

TIGER. Yeah. They're very good them ties. They made 'em good then.

GEORGINA. Ever the diplomat.

TIGER. Why?

GEORGINA. Who knows? He was never really himself again after the fire.

TIGER. And her ladyship?

MILES. Mummy died soon after. Of a broken heart.

GEORGINA. Mummy died of a clapped-out liver.

MILES. That's it, do her down!

GEORGINA. Then don't imply Daddy was barmy.

TIGER. 'E wasn't barmy, not the colonel.

Pause.

COOP. Spider will be wondering, Tiger. We left him in charge. We'd better go.

GEORGINA. So soon?

COOP. There's a lot to do. Pitch the tents, make a kitchen.

MILES. Dig the latrines.

GEORGINA. Not at teatime Miles.

MILES. Can't ignore the realities, can we Coop?

COOP. It is unwise.

MILES. I say, here's a wheeze! How would it be if we came to visit you tomorrow? When you've set up camp? (*Slight pause.*) Tiger?

TIGER. Mmm? Oh, yeah, course you can Master Miles, it'd be a pleasure.

MILES. Whooppee! Going camping again!

TIGER. And Miss Georgina, p'raps you'd care to 'onour us.

GEORGINA. I don't know Tiger. Got any gin?

TIGER. Now, now, Miss Georgina.

GEORGINA. Yes, why not? It'll be fascinating to watch you strapping lads scouting away. Until tomorrow then, Mr Cooper.

COOP. Thank you for the tea.

TIGER. It's — it's good to see you both again, you know.

GEORGINA. It's good to see you, Tiger.

MILES. Rather. You haven't changed a bit.

TIGER. An' I'm sorry to 'ear about the colonel.

MILES. Bit of a shock for you, I daresay.

TIGER. Yeah well. Thankin' you for your generous 'ospitality. Honoured to return the invitation.

MILES. Delighted to accept, old chap. And if there's anything you want, you just let us know.

TIGER. Well, there is just one thing, if I might make so bold.

MILES. What's that?

TIGER. Can I 'ave another of them little pink cakes?

Cut to the camp fire. It is night; owls call. The scene is subdued.

SPIDER. Everyone present and correct, Tiger.

TIGER. And the new lads?

SPIDER. Fast asleep. I told them they've got to be ready for the important visitors tomorrow. I say, aren't you turning in?

TIGER. Me and Coop fancies sleepin' out by the fire. It's a lovely night.

SPIDER. Bob's your uncle. Good night, then.

COOP. ⎫
TIGER. ⎭ Night, Spider.

SPIDER. I'm glad we're staying. We like it here. (*He goes.*)

COOP. Well, time to bed down.

TIGER. Shan't be a jiff. Finish me cocoa. (*He slurps.*) 'Ere, aren't you gonna shake out your sleeping bag?

COOP. What for?

TIGER. "You will always see an old hand in snake country look through his blankets before 'e turns in."

COOP. Not many snakes up the old A1.

TIGER. Don't get clever, Coop, it don't suit you. (*He slurps.*) What a day though, eh? You know, it sort of feels like I've never bin away. And when I done the servin', well. I 'adn't lost me touch. Bravo they said.

COOP (*smiles*). You aren't half an old hypocrite, you know.

TIGER. 'Oo? Me?

COOP. Scented muck you called it. *And* I saw you sneaking the boot-blacking sandwiches.

TIGER. Well, they was our 'osts. You gotta be polite. I was real proud to show you off, you know that? And I could tell they liked you.

COOP. Oh?

TIGER. Don't think I didn't see Miss Georgina givin' you the glad eye!

COOP. Oh, come on!

TIGER. Well, you're a personable lad. And she's a woman.

COOP. Of sorts.

TIGER (*sharply*). 'Ere. "A Scout is polite to all — but especially to women and old people and invalids, cripples etc."

COOP. What does she come under?

TIGER. I concede she ain't no oil paintin'.

COOP. She's no miniature, either.

TIGER. No, that's true. But she's got a way with her. Take my word for it.

Cut to TIGER's *pantry.*

YOUNG GEORGINA. Play with me Tiger. Play 'This is the way the lady rides.'

TIGER. 'Ere, now look. You get off.

YOUNG GEORGINA. Up and down, up and down.

TIGER. That's enough o' that. You should be at your lessons. (*A door opens.*) Oh gawd!

YOUNG GEORGINA. Hello Daddy.

COLONEL. Timms!

TIGER. It ain't what you're thinkin'.

COLONEL. You pervert! Horny hands on delicate buttocks!

TIGER. Delicate?

COLONEL. Dandling my daughter!

TIGER. Dandlin'? You seen the size of 'er?

YOUNG GEORGINA. Beast!

COLONEL. I'm beginning to rue the day I brought you here, Timms.

TIGER. Never mind rue. You wants to be thankful.

COLONEL. Thankful?

TIGER. Yeah. I could've been bent. It could've bin Master Miles.

Cut back to the camp at night.

COOP. And he's a bit unusual too. Her brother. All that 'whizzo' and 'what a wheeze'. Is he retarded?

TIGER. Bit deaf, that's all.

COOP (*yawning*). A blow received in childhood.

TIGER. All right, I can take a hint. Shan't be a tick.

COOP (*sighs*). Now what?

TIGER. "I have often used my boots or shoes as a pillow rolled up in a coat so that they don't slip apart."

COOP. Ah, so that's why your head's a funny shape?

TIGER (*laughs, a sort of cackle*). That's good that. (*He settles down.*) I likes you, Coop.

COOP. I like you, Tiger.

TIGER. We gets on.

COOP. Yeah.

TIGER. Give us a kiss.

COOP (*smiling*). Go to sleep.

TIGER. Ony jokin' boy, ony jokin'. Lovely night, ain't it? Look at them stars. Plough, Orion's Belt. I was able to teach 'im the stars, you know. On account of servin' under the Red Duster. We was mates, you know, like you an' me. (*Pause.*) I wonder why 'e done it, eh? 'Ow would you feel if I done what 'e did? Coop? Boy? Good night, son.

Cut acoustic.

TIGER (*internal*). "Camp Fire Yarn Number Three. How to Grow Strong. If you want to get through adventures safely when you are a man and not be a slopper, you must train yourself to be strong, healthy and active as a lad. The secret of keeping well and healthy is to keep your blood clean. A short go of exercises every morning is a grand thing for keeping you fit."

Fade up campsite in the early morning with the Scouts assembled.

TIGER. Patrol, attention! Now, Scouts, we 'ave bin 'ighly 'onoured by the visit this mornin' of Miss Georgina and Master Miles. They 'as generously allowed us to set up camp in this delightful spot for what I am sure will be a memorable scoutin' week-end. Let us show our appreciation in the usual manner. Hip hip!

ALL. Hurray!

TIGER. Hip, hip!

ALL. Hurray!

TIGER. Hip, hip!

ALL. Hurray!

GEORGINA. Tiger, my head!

TIGER. Wossup?

GEORGINA. I'm afraid I'm rather in need of my morning medicine.

TIGER. "The best medicine is open air and exercise and a big cup of water in the morning if you are constipated."

GEORGINA. Fascinating.

COOP. We're ready for the demonstration, Tiger.

GEORGINA. Demonstration?

COOP. Over to you, Spider.

SPIDER. Thanks. (*He coughs.*) "Here are some good exercises to work up the circulation of the blood. Remember, the heart is the most important organ in a lad's body."

MILES *sniggers.*

GEORGINA. Quiet, Miles. (*She sniggers.*)

SPIDER. "Raise the hands gradually over the head and lean back as far as possible drawing in a deep breath as you do. While looking up in this way say to God, 'I am yours from top to toe.' " Ready? and raise —

ALL. 'I am yours from top to toe.'

SPIDER. "Lower the arms gradually, breathing out the words 'Thanks to God.' " And —

ALL. "Thanks to God."

SPIDER. Repeat.

They continue the exercise and chanting.

GEORGINA. Very impressive, Tiger.

TIGER. Good, ent they?

GEORGINA. Very Nuremburg.

MILES. Takes me back!

COOP. Brownsea Island, actually.

GEORGINA. I'm sorry, Mr Cooper?

COOP. They had different uniforms at Nuremberg.

GEORGINA. Really? Anyway, I don't know how you can all be so
jolly this early in the morning. It's obscene!

TIGER. Call this early? We bin up since six!

MILES. Golly!

TIGER. 'Ad breakfast by seven.

MILES. Don't tell me! A pint of tea and —

TIGER. A plate of prunes! You remembered!
All right Spider, that's enough.

SPIDER. Scouts, attention!

GEORGINA. ⎫
MILES.   ⎬ Bravo, well done etc. (*They clap.*)

TIGER. Now remember: "Doctors say that half the good of exercise is
lost if you do not have a bath immediately after it." And we've got
a topping bathing place, ain't we, thanks to the lady and gentleman,
so — off with the togs and last one in's a cissy!

The SCOUTS *cheer and laugh, splashing sounds.*

MILES. I say, there's a sight.

GEORGINA. Not taking your togs off Mr Cooper?

COOP. I bathed earlier thank you.

TIGER. 'Appy as Larry, ain't they?

MILES. I must say you've got it looking jolly smart, Tiger.

TIGER. "The camp ground should at all times be kept clean and tidy."

GEORGINA. And I envy you your kitchen area.

MILES. I could eat one of your stews now Tiger. Yum, yum!

TIGER. But you got a sparklin' modern kitchen up there.

GEORGINA. My dear Tiger, even you must have noticed that the old
home isn't quite as stately as once it was.

TIGER. Nothin' that a good spring clean and a flood of sunshine
won't put right.

GEORGINA. The sunshine we can afford.

TIGER. Like that is it?

GEORGINA. It is rather. (*Beat.*) Well, Miles. Work to be done —

TIGER (*with an air of decision*). 'Ang on. I got an idea.

GEORGINA (*innocently*). Yes?

TIGER. Well, we're down 'ere, a body of Scouts willin' and eager, and you're up there in need of 'elp —

GEORGINA. Oh we couldn't possibly —

TIGER. Now, listen. "When you Scouts use a farmer's ground you ought to repay him or her in some way." Well, it's tailor made, ain't it?

GEORGINA. But what about your camping and adventure?

COOP. Yes — what about them, Tiger?

TIGER. Woss the odds? It's all scoutin'. You rely on us. We'll go through the place like a dose of salts.

GEORGINA. Well, I'm overwhelmed. Thank you. And thank you, Mr Cooper.

MILES. I say, and after we'll have a party. Rabbit pie and lemonade!

TIGER. A musical evenin'. Like the old days! Right, we'll bring the lads along later.

MILES. I must say they look a pretty efficient bunch.

GEORGINA. Strapping chaps.

TIGER. They are. I keeps them in tip-top condition.

Cut acoustic.

TIGER (*internal*). "So practise keeping healthy and then you will be able to show others how to keep themselves healthy too. In this way you can do many good turns. When you get up in the morning, remember that you have to do a good turn for someone during the day. You must do one. But if you can do fifty so much the better."

COLONEL. Timms, I pay you to keep this place clean. Get on with it.

TIGER. What you want with a house this size anyroad? You could get twenty 'omeless families in 'ere.

COLONEL. Ah, the politics of envy.

TIGER. Property is theft, mate.

COLONEL. Arseholes.

TIGER. Well, that's charmin' ent it? 'Course it's all right for you and

Baden-Powell. Stuck at the top of the bleedin' tree.

COLONEL. You know what I can't stand about your half-baked Marxist clap-trap? There's no joy. No one smiles.

TIGER. The workin' man ain't got a lot to smile at, mate.

COLONEL. Get on with your work. With a smile, Timms. And a cheery whistle. If you please.

Cut acoustic. Fade up 'Ging Gang' whistled in the background. Hold under the following.

TIGER. Right, Spider, you and your men, you're in charge of the hupstairs. Corridors, bedrooms, the usual offices. Coop, you take Tucker and Pringle and the other new ones, houtside. You're i.c. the grounds, the gardens and the messuages. Me an' my lot, downstairs. Any questions? Right, let's set to. 'Old it! Tucker, what's the rule about shirtsleeves?

TUCKER. Sorry, sir. "All Scouts should have them rolled up because this tends to give them greater freedom."

TIGER. Good lad. Pringle?

PRINGLE. "And also as a sign that they are ready to carry out their motto."

TIGER. Which is?

ALL. Be Prepared!

Cut whistling dead. Fade up musical box and cut. GEORGINA *and* MILES *close together, the sounds of working in the distance.*

MILES. They're working jolly hard.

GEORGINA. Isn't that what they're here for?

MILES. Funny, having old Tiger back.

GEORGINA. Mmm. Oh, did you find his present?

MILES. In his old attic bedroom. I wrapped it up in its box. I suppose he will like it?

GEORGINA. Course he will.

MILES (*listens*). I think we ought to lend a hand.

GEORGINA. Nonsense. I won't let you go.

MILES. Oh, get off!

GEORGINA. Careful — he bites!

MILES. Well — don't think I didn't see you looking at him.

GEORGINA. Who?

MILES. That Patrol Leader chappie.

GEORGINA. Don't be a silly! What would I do without old Milo, eh?

MILES. Huh.

A dinner gong is heard, struck vigorously.

GEORGINA. Luncheon is served, by the sound of it.

MILES. We haven't used the gong for years, it's the dickens of a row.

GEORGINA. What d'you expect from him, subtlety?

MILES. Who?

GEORGINA. Dear Miles, there's only ever been one person who banged the gong so hard that even you could hear it.

Bring up the gong very loud and finish.

COLONEL. For God's sake, Timms!

TIGER. Look, I'm a butler, and this is a gong. Butlers 'it gongs.

COLONEL. But we're all assembled!

TIGER. 'Cept 'er. Where is she?

COLONEL. 'My wife' will be taking breakfast in her room.

TIGER. Them bleedin' stairs.

YOUNG MILES. Daddy, he swore again!

COLONEL. Just bring the food, Timms.

YOUNG GEORGINA. He called me a little tart yesterday. What's a tart, Daddy?

TIGER. 'Ere you are. I 'ope it chokes you.

COLONEL. What the devil's this?

TIGER. Breakfast.

COLONEL. Timms — I take bacon in the mornings, the children have prunes, and my wife likes an apple from the orchard and honey from the hive.

TIGER. For a start, I 'ad the last of the pruins; second, it's pissin' down outside; and third, I'm fed up of bein' stung.

COLONEL. But black pudding!

YOUNG MILES. It looks horrid.

TIGER. 'Olesome, that. Made out of pig's blood.

YOUNG GEORGINA. ⎫ Ugh!
YOUNG MILES.   ⎭

TIGER. Oh, and 'ere's yer paper.

COLONEL. Where's *The Times*?

TIGER. I cancelled that. Thought it's time you read the truth.

COLONEL. *The Daily Worker*? Right, this communist drivel is going straight on the fire!

TIGER. Right, that's it, I resign.

YOUNG MILES. Hurrah!

YOUNG GEORGINA. Good riddance!

COLONEL. Oh, God, not again.

TIGER. An' if you wants me, you knows where to find me!

Cut to the kitchen and the sound of a knife and fork being put on a plate.

TIGER. Lovely grub. (*He belches.*) Beg pardon. Spider's cookin's improvin', ain't it?

COOP. So this was your domain, was it?

TIGER. Yup. This was the engine room of the 'ole concern. You could always find Timms in the kitchen.

COOP. I thought butlers had pantries.

TIGER. They do. That was mine over there in the corner.

COOP. The green baize door. Pity it's blocked off.

TIGER. Yeah . . . an' I seen this table piled 'igh you know. Partridge, pheasants, rabbits and hares. Vegetables from the kitchen garden. Fruit from the orchard. And fish: caught a pike once, we did, me an' 'im, down at the pool. Great teeth it 'ad. 'Ow you goin' on outside?

COOP. It'll be some time before the kitchen garden's back in production.

TIGER. Attractive grounds though, ain't they?

COOP. You could build a few housing estates out there.

The door opens.

SPIDER. Tea up!

TIGER. Just what the doctor ordered.

COOP. Thanks Spider. (*He takes the tea.*)

SPIDER. Was the dinner all right?

TIGER. Lovely 'Bannocks, bacon and beans.' Can't beat it.

SPIDER. Mister Miles had three lots.

TIGER. And Miss Georgina?

SPIDER. Well, she had to keep taking medicine from a bottle, but she did try some.

COOP. You've made a bit of a hit there, Spider.

SPIDER. I think they're a very nice lady and gentleman. Oh, message for you Coop, from the lady. She wants to see you after dinner.

TIGER. Miss Georgina?

SPIDER. In private.

COOP. You an errand boy now, Spider?

SPIDER. Only doing a good turn.

COOP. What does she give you?

SPIDER. Eh?

COOP. For running messages.

SPIDER. A Cox's Orange Pippin, actually. From the orchard. Scrumptious. Cheerio.

He exits, whistling 'Ging-Gang'. TIGER *chuckles quietly.*

TIGER (*cackles*). In private, eh?

COOP. Lay off.

TIGER. I think it's lovely. My ol' Coop an' her. Make an old man very 'appy.

COOP. "Do not spend time on a girl whom you would not like your mother or sister to see you with."

TIGER (*teasing*). That ain't very courteous. Ain't you goin' then?

COOP. I might. When I've finished my work, I might.

TIGER. Son, when you're called, you 'as to go.

Cut to flashback. TIGER *in the pantry singing,* COLONEL *outside banging on the door.*

TIGER (*behind the door, drunkenly singing to the tune of 'John*

Brown's Body ').

'They'll make Sir Winston Churchill smoke a Woodbine cigarette
When the Red Revolution comes.'

COLONEL (*bangs*). Timms, that's my best Port!

TIGER. Hexcellent vintage the '22.

COLONEL. It's priceless. I keep it for guests.

TIGER. Let them drink brown ale! (*He cackles.*)

COLONEL. I can't give the Lord Lieutenant of the county brown ale!

TIGER. Oh, 'e's scroungin' another 'ot dinner, is 'e?

COLONEL. You can't resign now!

TIGER. Serve 'im yourself.

COLONEL. Bugger it man, what do you want?

TIGER. Blood.

COLONEL. I'll raise your wages. Five pounds ten.

TIGER. More blood!

COLONEL. Six pounds.

TIGER. You don't know where you are without me, do you comrade?
I ain't wearing that penguin suit no more.

COLONEL. You'll stay?

TIGER. Say please.

COLONEL. Please.

TIGER. More.

COLONEL (*sighs*). Please, comrade.

TIGER. That's better.

> *Cut to the musical box in the nursery.* GEORGINA *hums along.*
> *There is a knock on the door.*

GEORGINA. Come. (*The door opens.*) Why, Mr Cooper. You needn't
have knocked. No formalities here.

COOP. What do you want?

GEORGINA. My, my, straight to the point. (*She switches off the
music box.*) A musical box is such a charming toy, don't you think?

COOP. We didn't have one in our nursery.

GEORGINA. Of course. I'm sorry, one forgets. I'm very fond of this room. So many memories. When we were little, Tiger used to lock us in if we were naughty. 'Off to the nursery,' he'd shout. Mind you, we did rather give him hell.

COOP. Poor old Tiger.

GEORGINA. Not that he didn't deserve it.

COOP. Look, I want to get cleaned up. I've been finishing the garden.

GEORGINA. You know, we do so appreciate what you're doing for us.

COOP. Sooner it's done, the sooner we can get back to camp.

GEORGINA. Oh dear, you are restless. What can I do to keep you? Perhaps you'd like to play with our toys. There's my doll's house, or Miles's bow and arrow. The train set. Or would you like a ride?

COOP (*beat*). I'm sorry?

GEORGINA. Foxhunter's very docile. He never throws you.

COOP (*evenly*). I don't ride.

GEORGINA. Then perhaps you'll give me a hand. (*Beat.*) To mount.

COOP. D'you need a hand?

GEORGINA. 'Fraid so. I believe I am what is commonly known as a 'Big Girl'.

COOP. What?

GEORGINA. Good breeding stock, child-bearing hips. (*She laughs.*) Oh come on, don't be a spoilsport. Help me.

He helps her onto the rocking horse.

Thanks. I like a good gallop. If you pull the rein, he'll move. Steady! Gently. That's it. Up and down, up and down.

The horse jingles and creaks.

COOP. Tiger will be expecting me.

GEORGINA. Where did you meet him?

COOP. Can't you guess?

GEORGINA. I'm sorry, you were there as well?

COOP. Not exactly. I visited.

GEORGINA. I think you've been an improving influence on him, Mr Cooper. He's a changed man.

COOP. He's a good friend.

GEORGINA. That's what Daddy thought. Up and down. I don't want to tell tales out of school —

COOP. Then don't.

GEORGINA. But Tiger wasn't a very . . . committed Scout in the old
days. He used to cross out all the queen and country bits from
Baden-Powell. And for bedtime stories, we had readings from
Das Kapital.

COOP. I'm horrified.

GEORGINA. And he used to swear, can you imagine? Swear
dreadfully. And, I'm afraid, he hit the port rather. (*Beat*.) Look, I'm
only telling you this because I want you to realise that whatever
happened to him — well , he's ever so much improved.

COOP. Thank you for your concern. Now if you'll excuse me.

GEORGINA. I see I can delay no longer. Whoa, boy! (*The horse stops*.)
Mr Cooper, what do you do? Apart from night school and visiting?

COOP. I haven't got anything permanent at present.

GEORGINA. I see. (*Beat*.) Look, I'll come to the point. I like you. This
is a big house and estate. An extra pair of hands would be invaluable.

COOP (*taken aback*). I beg your pardon?

GEORGINA. I could offer you a very interesting . . . position.

COOP. You mean a job?

GEORGINA. That too. Nothing onerous. You'd have plenty of free
time for your studies. I'd want only the occasional . . . conversation.

COOP. What about your brother?

GEORGINA. Oh, Miles shares my feelings.

COOP. Is that all he shares?

GEORGINA. What *can* you mean?

COOP. "One of the most important things a Scout has to learn is to let
nothing escape his attention."

GEORGINA. Yes, well, passing lightly over that, have you ever seen a
dead body?

COOP. What?

GEORGINA. "It may happen to you Scouts that you will be the first
to find the body of a dead man." Be warned, it is not pleasant.
Especially when it's your own father.

Pause.

COOP. Look, I'm not butler material.

GEORGINA. Neither was Tiger. The terms would be generous. Think about it, Mr Cooper. Are you sure you wouldn't like to . . . serve me?

The door opens.

MILES. Georgie, here you are. I've got the — (*He sees Coop.*) Oh.

GEORGINA (*sighs*). Hello, Miles.

MILES. Well, this is very cosy.

GEORGINA. Now darling —

MILES. They were looking for you downstairs Mr Cooper. Everyone's ready.

COOP. Ready?

MILES. For the party of course.

GEORGINA. Oh, you can't miss the party.

COOP. I'd better go and help.

MILES. It's all done. That bright young chap, what's his name, Spider, he's *jolly* efficient.

COOP. I'm sure he was.

GEORGINA. Don't sulk, Miles, or you shan't have your treat.

MILES. Treat?

GEORGINA. Well, as it's a special occasion, I thought I'd unlock the Tantalus.

MILES. Oh Georgie, would you?

GEORGINA. If you're good. A celebration.

MILES. Whizzo!

GEORGINA. We've had some parties here, Mr Cooper.

MILES. I'll say, old Tiger the life and soul!

GEORGINA. So let's make this one to remember.

Cut acoustic.

TIGER (*internal*). "Camp Fire Yarn Number Four. Want of laughter means want of health. If you read about Scouts you will generally find that they are pretty cheery old fellows, especially at Camp Fire, which is one of the happiest hours of camp. Songs, recitations and small plays can be performed round the Camp Fire and every Scout should be made to contribute something to the programme whether he thinks he is a performer or not."

*Cut to complete chaos and uproar in the dining-room, children
shouting, dog barking etc. Tiger is singing the 'Hokey Cokey' loudly
and drunkenly.*

TIGER. 'Oh hokey cokey cokey. Oh hokey' *etc.*

COLONEL. Timms, get off the table you maniac!

TIGER. Everyone does a turn round the camp fire. What your book
says. 'Oh hokey cokey cokey' *etc. etc.*

COLONEL. This is a dinner party and the Lord Lieutenant does not
want a cabaret!

TIGER. 'Ello squire. Enjoying' the show?

COLONEL. I am dreadfully sorry, sir.

YOUNG GEORGINA. Daddy, Daddy, Tiger's taking his clothes off.

COLONEL. Oh my God.

TIGER. 'You puts yer left leg in . . .'

COLONEL. Put your clothes on at once. If I catch you —

YOUNG MILES. Daddy, the lady has fainted!

TIGER. Wossup, ain't she seen a real man before? 'Ere darlin'. 'In, aht,
in aht, shake it all about.' (*Cackle.*)

COLONEL. Turn your faces to the wall, darlings. No, my lord, please
don't go!

TIGER. 'You puts the 'ole lot in, the 'ole lot aht.'

COLONEL. Sing children, sing with me! (*He sings:*) 'On the road to
Mandalay, Where the old flotilla lay.'

TIGER. Imperialist crap! 'Oh, hokey cokey.'

COLONEL. ⎫
GEORGINA. ⎬ 'Can't you hear their paddles chunkin' from Rangoon
MILES. ⎭   to Mandalay.'

*Dead cut to the present. TIGER and MILES singing 'Mandalay'
lustily, picking up the song but a tone higher.*

TIGER. ⎫'On the road to Mandalay, Where the flying fishes play And the
MILES. ⎭dawn comes up like thunder Out of China 'cross the bay!'

They make a big finish to cheers, applause, laughter etc.

MILES. Well done Tiger!

TIGER. Thass a good old song. Haven't sung that in years!

MILES. Rather. Well, who's next, eh? I could do card tricks. Find the lady, penny stakes?

GEORGINA. Ah, Mr Cooper, aren't you going to do a turn?

COOP. I'm not very good at that sort of thing.

SPIDER. I am. Here's another tongue twister.

TIGER. Yeah, all right lad —

SPIDER. 'Are you copper bottoming 'em, my man?
No, I'm aluminiuming 'em, Mum.'

MILES. Gosh. 'Are you copper bottomomingingalom.'

Laughter.

GEORGINA. Useless, Miles.

SPIDER. It's easy. 'Are you copper bottoming —'

TIGER. That's enough, thanks Spider. In fact, I think it's time we
was making tracks.

General groans.

GEORGINA. Oh, you can't go yet.

COOP. There's a lot to catch up on tomorrow. Need an early start.

TIGER. It's bin a really toppin' party ent it lads?

ALL. Yes!

GEORGINA. But we've still got the high spot of the evening to come.
So I insist we all have one more little drinkie. Miles.

MILES. Righto. Thirsty work, this singing.

TIGER. 'Igh spot?

SPIDER. I'll take the lemonade round.

MILES (*whispers*). D'you want something a bit stronger, Coop?

COOP. I don't drink, thanks.

GEORGINA. I think I need a ciggie. Got a match, Tiger?

TIGER. I — er — don't carry 'em.

GEORGINA. How odd.

SPIDER. I do. Here, Miss.

MILES. Right. Everyone got a drink? Good. Here's a toast then. (*He
clears his throat.*) To you Tiger and to the Curlew Patrol. Simply,
thanks.

GEORGINA. Yes. An absolutely splendid effort. The house looks
like new.

TIGER. It's us should be thankin' you. For lettin' us stay and the lovely dinner an' all.

GEORGINA. So as a token of our thanks, we've got a little presentation for you. Now everyone gather round.

General movement and murmurs.

TIGER. Presentation?

GEORGINA. Have you got it, Miles?

MILES. Here. All wrapped up.

GEORGINA. Good. Tiger — step forward!

TIGER. Me? I don't deserve nothin'.

MILES. Tiger, in appreciation of all the hard work you've done today —

GEORGINA — and as a thank you for all the happy memories you've revived —

MILES. We'd like you to accept this.

The box is handed over.

GEORGINA. Go on, open it up, Tiger.

TIGER. What is it?

The box is unwrapped.

(*Softly:*) Well, I'm blowed.

SPIDER. What is it, Tiger? Show us!

TIGER. All these years, you kept it.

GEORGINA. We thought it might come in useful one day.

SPIDER. ⎱
ALL.　⎰ (*ad lib*). Show us, Tiger! Let's have a look!! (*etc. etc.*)

TIGER. It's my old uniform, lads. When I was butler.

ALL (*ad lib reaction*). Cor! Gosh!

MILES. The old penguin suit eh?

TIGER. Jest 'ave a look at that will you, Coop?

COOP. Very nice.

TIGER. I — I don't know what to say.

MILES. No need, old chap.

GEORGINA. Well, now Tiger. Going to try it on?

TIGER. Try it on? Now?

GEORGINA. It's what it's for.

TIGER. Well, I don't know. What d'you reckon, lads?

ALL. Yes!

TIGER. Coop?

COOP. Up to you.

TIGER. Miss Georgina, Master Miles. It will be a privilege — and an honour.

Cut to the sound of jacket being vigorously brushed.

COLONEL. Go easy with the brush, man!

TIGER. Stop whinin'. You wants the nap raised, don't you?

COLONEL. And perhaps my boots need another shine. Got to look my best for Rememberance Day parade.

TIGER. The massed ranks of hypocrisy.

COLONEL. What?

TIGER. All them time-servers and desk job johnnies mournin' men like me, blown to bits in the mud.

COLONEL. You? You never left Aldershot.

TIGER. Ha bloody ha. Turn round.

COLONEL. Well, how do I look?

TIGER. Hm. (*Slight pause.*) I think you look ridiculous.

COLONEL. The full dress uniform of a colonel in Her Majesty's forces, ridiculous?

TIGER. All that blue and gold. Feathers in yer 'at. You look like an old parrot.

COLONEL. Listen, you ungrateful bastard. "My country and your country did not grow of itself out of nothing. It was made by men and women by dint of hard work and hard fighting often at the sacrifice of their lives — that is, by their whole-hearted patriotism!"

TIGER. That bloody book.

COLONEL. And this uniform is a symbol of that patriotism!

TIGER. Sod patriotism — and sod your uniforms.

Cut back to the party. General murmur

MILES (*slightly off*). Nearly ready, everbody! He looks terrific!

GEORGINA. Anyone got another turn while we're waiting?

SPIDER. I know a monologue Tiger taught me. (*He coughs.*) 'There's a little yellow idol to the —'

GEORGINA. Yes, thank you dear. Is there nobody else?

COOP (*quietly*). I know a song.

GEORGINA. Why, Mr Cooper!

SPIDER. Is it that one you learned at night school?

COOP. I haven't much of a voice.

GEORGINA. No matter. Don't be shy. (*Beat.*) Well, Mr Cooper, we're all ears.

COOP (*softly sings*). The people's flag is deepest red
It shrouded off our martyred dead
And ere their limbs grew stiff and cold
Its heart's blood dyed its every fold;
Then raise the scarlet standard high,
Within its shade we'll live or die —

GEORGINA. The working class can kiss my arse
I've got the foreman's job at last.

COOP. I don't think that's very funny.

GEORGINA. Don't you? I think it's a hoot.

MILES (*slightly off*). OK everybody, we're ready.

GEORGINA. Isn't this exciting?

MILES. Somebody ring the bell. That sash by the fireplace.

SPIDER. I'll do it! I'm good at that sort of thing.

COOP. Spider —

GEORGINA. Jolly good. Ready then? Pull!

The bell rings in the distance. Suppressed excitement. The door opens and there are footsteps on the marble floor. They approach and stop. Silence.

TIGER. You rang, madam?

Applause, calls of bravo and cheers.

Thank you, thank you very much, thank you. (*Applause dies.*) Well, 'ow do I look? Eh, Spider.

SPIDER. I haven't seen you out of Scout togs before. You look . . . very nice.

TIGER. Ta. Master Miles?

MILES. Top hole! Wizard. The old Tiger.

TIGER. Miss Georgina?

GEORGINA. I think you look . . . thoroughly at home.

TIGER. Suits me, don't it? — Coop?

COOP (*slight pause*). Spider, get the lads together and get them outside. We're going back.

SPIDER. Eh?

COOP. You heard.

TIGER. I said, 'ow do I look, Coop?

Pause. Silence.

TIGER. Well, I know I put on a bit of the old avoyer dupoyer an' the trousers are a bit short. But come on. Honest opinion.

COOP (*beat, quietly*). I think you look ridiculous.

Dead cut to flashback.

COLONEL. Timms, I have had as much as I can stand of your rank insubordination. I give you a chance in life where no one else will —

TIGER. Oh very 'umble —

COLONEL. Quiet! I put up with it — even when you dance naked before my guests — because I believe you might be worth it. But no. Now you insult all I stand for. Deride all that is decent and good. My monarch. My country. My class and my profession. The hand that fed you.

TIGER. Fed?

COLONEL. There's no room for you here, Timms.

TIGER. Don't talk daft.

COLONEL. You lack charity, humility and any sense of social responsibility. You are dismissed!

TIGER. Eh?

COLONEL. You heard. Fired!

TIGER. Right. Fired is it? Right.

COLONEL. Out, Timms!

TIGER. Arter all I done. We'll see about fired!

COLONEL. *Out*!

Cut to interior of tent. TIGER *is reading aloud quietly from* 'Scouting for Boys'. *He reads a little haltingly.*

TIGER. "Camp Fire Yarn Number Five. Citizenship. A camp is a roomy place. But there is no room in it for the fellow who does not want to take his share of the many little jobs that have to be done."

COOP. I'm sorry.

TIGER (*beat*). "There is no room for the shirker or the grouser — well, there is no room for them in the Boy Scouts Movement at all, but least of all in camp."

COOP. I didn't like to see you humiliated.

TIGER. "A Scout must never be a SNOB." 'Oo was·it done the 'umiliating? "A Scout accepts the other man as he finds him and makes the best of him. In this way comradeship grows —"

COOP. Oh for God's sake! (*He snatches the book.*)

TIGER. Gimme my book!

COOP. I'm trying to talk to you!

TIGER. I don't want to know. Now 'and it over.

COOP. Listen, will you?

TIGER. Ridickerlus you called me, in front of everyone.

COOP. I was trying to bring you to your senses.

TIGER. Nothin' wrong with my senses. I can't make you out. Perfectly nice people give me a present. What's ridickerlus about that?

COOP. Nice people? Look at you, lying there in a joke butler's uniform. Nice people?

TIGER. Well, 'e's fond of a prank, Master Miles. Always was.

COOP. Running off into the night with your Scout uniform and her passing out on the floor with laughter, a prank?

TIGER (*beat*). Give me me book.

COOP. And as for having an army of servants for the day and then offering me your old job, — what a hoot!

TIGER. Offerin' you butler?

COOP. That's why she wanted to see me.

TIGER. Cor! What you say?

COOP. What d'you think I said? You don't know me at all, do you?

TIGER. Course I does. My best mate. But I don't know what's got into you. (*Beat.*) Look, forgive and forget, he'll bring me togs back, you'll see. I mean, they got no reason to make a fool of me.

COOP. What?! You're blocking it all out, aren't you?

Slight pause.

TIGER. Dunno what you mean.

COOP. Tiger — who gave you this book?

TIGER. You know 'oo give it me.

COOP. And what happened to him, their daddy?

TIGER. Leave it out.

COOP. OK, you want to read it. Here it is.

TIGER. What?

COOP. Go on, take it. Mustn't strain your eyes though. I'll light another lamp.

TIGER. Now steady on, Coop.

COOP. 'Course, we shouldn't really have paraffin lamps in a tent, bit dangerous. Where's my matches?

TIGER. Stop mucking about!

COOP. Steady Tiger, you'll knock it over.

TIGER. That match, you'll 'ave the place alight!

COOP. What happened, Tiger?

TIGER. Leave me alone.

COOP. What happened?

TIGER. Put it out. Please! Put it out!

Cut to the COLONEL *banging on the pantry door.* TIGER, *slightly drunk, is on the other side.*

COLONEL. Open this door, you madman!

TIGER. 'Oo you callin' mad?

COLONEL (*sniffs*). Paraffin! Christ, open this door.

TIGER. Fired you said, fired you shall have!

COLONEL. I take it back!

TIGER. Too late. And now the doomed man will 'ave a final cigarette. 'E strikes the match, so. What a pretty flame. (*Beat.*) Oh dear.

COLONEL. What's happened?

TIGER. I dropped it. Still, the West Wing'll look lovely all lit up.

COLONEL. You'll burn to death in there!

TIGER. We shall all burn! It's the only way. (*He sings.*) 'The people's flag —'

COLONEL. I'll give you anything.

TIGER. Blood, I want. Rivers of blood. (*He sings.*)

COLONEL. You blasted commie. Queen and country!

TIGER. Rivers of blood. (*He continues singing.*)

COLONEL. Right! (*He sings.*) 'And did those feet In ancient times, walk upon England's mountains green.'

They sing loudly against each other. The flames crescendo and fade out. Back to the tent. TIGER *is upset.*

TIGER. I'm an old man, Coop. Leave me alone. It was a wrong thing I done, a wicked thing. Why d'you make me remember?

COOP. No, it was — a beautiful thing.

TIGER. Oh, the young zealot! You got it wrong lad. It was malice. I was in drink. See, Coop . . . forgive me. I used to be . . . a drunken man.

COOP. I know.

TIGER. Eh? But you — you never said.

COOP. Forgive you said. Which was more than he did.

TIGER. Let me tell you, lad, he got me out of that pantry. I don't know 'ow 'e done it. See 'is 'ands now. All bleedin'.

COOP. But you didn't want to be got out.

TIGER. Yeah, I 'ad some half-baked notion of bein' a martyr, goin' up in flames for the cause.

COOP. All or nothing. Perhaps it is the only way.

TIGER. But it *ain't* lad. That beautiful 'ouse.

COOP. Bugger the house!

TIGER. Cooper!

COOP. I'd like to see the whole lot a pile of rubble!

TIGER. You would, would you?

COOP. All or nothing.

TIGER. 'Ere you are then, 'ere's yer matches.

Sound effect: a box of matches.

COOP. What?

TIGER. Go on, up the hill. You knows where it is. Go an' finish
the job.

COOP. Don't talk daft.

TIGER. Ah, ain't that easy is it? Look, lad, I've stood where you're
standin'. Everythin' black and white. But you got to learn to get
along with folk. To see the best in everyone. Look, give me the
book. (*He rustles the pages:*) 'Ere, now listen. "We are very much
like bricks in a wall, we each have our place though it may seem a
small one in so big a wall. But if one brick crumbles or slips out of
place, it begins to throw an undue strain on others, cracks appear
and the wall totters."

COOP. Give me that! Now you listen. (*Pages rustle.*) "It often happens
that when you are tramping alone through the bush you become
careless in noticing in what direction you are moving. You
frequently change direction to get round a fallen tree, or other
obstacle and having passed it do not take up exactly the same
direction again. A man's natural inclination somehow is to *keep
edging to his right* and the consequence is that when you think you
are going straight you are not really doing so at all." Think about
it. Good night.

TIGER. Oh very clever, very night school! Now I'm a fascist am I?
Look lad, I ain't angry, so why are you?

COOP. Because of some inhumane bastard who bunged you in − !

TIGER. Don't you call him that! I won't hear it! What do you know,
eh? You never met 'im. We was mates when you was still a snotty
little kid in short pants.

COOP. Don't call me snotty!

TIGER. I won't have a word against 'im.

MILES (*outside the tent, calls softly*). Tiger, Tiger.

TIGER. You want to watch that chip on your shoulder don't cripple
you lad.

MILES. Tiger.

TIGER. What? Yes? 'Ello?

MILES *makes the Curlew call.*

Stop muckin' about, Spider, it's late.

MILES. Tiger, it's me.

Rustle of tent flaps.

TIGER. Master Miles! What you doin' 'ere?

COOP. Oh God.

MILES. I know it's late. Only I heard voices. Can I come in?

TIGER. Course. Squeeze in.

MILES. Sorry if I'm interrupting. Yes, well — I won't stay. I just came to give you these back, Tiger.

TIGER. Eh?

MILES. Your Scout togs. Georgie didn't want me to, so I ran off with them. Sorry they're creased.

TIGER. Well, now, that's very considerate, Master Miles. Told you, didn't I?

COOP. Whizzo.

MILES. Sorry, Tiger, it wasn't a very good joke.

TIGER. Water under the bridge, lad.

MILES. Yes, well, I'll push off.

TIGER. 'Ang on, you needs an 'ot drink first. Coop, run down an' see if Spider's got any cocoa left.

COOP. Right. I could do with some air.

Rustle as COOP *goes. Small pause.*

TIGER. Don't mind 'im. He can be a bit of an 'othead.

MILES. Well, I don't blame him. You know, it's funny seeing you again in the old uniform. I suppose I thought — well, I thought things could be like they were.

TIGER. Well, praps it's best they ent, lad.

MILES. Perhaps. But sometimes I wander about the house or sit in the garden thinking, and I wonder what happened to it all.

TIGER. 'Ow d'you mean?

MILES. Oh, I don't know. What happened to *Children's Hour*, Tiger, and toasting muffins by the fire?

TIGER. I bust the toasting fork, that's what 'appened.

MILES. And Luck of the Legion and Uncle Mac, what happened to them?

TIGER. I dunno. Still there, ain't they?

MILES. No, they're gone Tiger. Picture book heroes. All gone.

TIGER. Not all. You ever thought of becomin' a Scout again? You was always keen.

MILES. Oh Tiger, it's too late.

TIGER. Nonsense. Join us!

MILES (*sadly*). I'd like to, but I'm afraid I'm on the slippery slope.

The rustle of tent flaps being opened.

SPIDER. Anyone at home?

MILES. Oh, hullo.

SPIDER. Brought your cocoa, Master Miles.

MILES. I say, thanks.

SPIDER. Message from Coop, skipper. Says he'll sleep out tonight.

TIGER (*beat.*) I see.

MILES. Mmm. Jolly good cocoa.

SPIDER. 'Promotes Restful Slumber'.

MILES. Mmm?

SPIDER. That's what it says on the tin. You want any more, Tiger?

TIGER. Er — no thanks lad.

MILES. Finish this, then I'll be off.

SPIDER. Oh — you going home?

MILES. Not exactly. Well, sometimes if Georgie gets a bit — well, sometimes I stay out. Make a bivouac or bed down in the church if it's wet.

TIGER. I won't hear of it. Old Spider'll sort you something out.

SPIDER. Rather. There's room in my tent.

MILES. Gosh, could I?

SPIDER. And in the morning you can join in the exercises and bathing and —

MILES. And breakfast by the fire. Yummy!

SPIDER. Right, I'll go and fix you up a sleeping bag.

A rustle as he goes.

MILES. I like him. I'm sorry about tonight. It all went a bit wrong. I — I know Georgie can be . . . difficult, but well —

TIGER. You was always close. You know, I reckon she took a shine to our young Coop, eh?

MILES. Yes. That's wh- (*Pause.*) Tiger.

TIGER. What?

MILES. Well, it's been lovely to see you, but under the circumstances — well, I expect there are a lot of other campsites.

TIGER. W-what d'you mean?

MILES. I mean, well, perhaps it would be better if you went away.

TIGER (*softly*). Have a care, lad.

MILES. Mm? Goodnight.

TIGER. Have a care.

Cut to the hall of the house. A slight echo.

COLONEL (*very subdued*). Here's your suitcase Timms.

TIGER. Suitcase? We goin' on our 'olidays then?

COLONEL. I've packed everything I think you'll need.

TIGER. An' what's that car doin' in the drive? That ain't our car. And where's your case?

COLONEL. We're not going on holiday. I thought it would be best if . . . if you went away for a while.

TIGER. Ent you comin' then?

COLONEL. For a rest. I'm sorry, Tiger, there's nothing I can do now.

TIGER. Tiger? You ent never called me that before.

COLONEL. You'll be well looked after.

TIGER. Sort of 'oliday camp, is it?

COLONEL. And I shall come to see you.

TIGER. Where's the nippers? Ain't they gonna see me off?

COLONEL. I thought it better —

TIGER. Well, tell 'em Tiger said ta-ta. I'll send 'em a picture postcard.

COLONEL. And my wife sends you her good wishes.

A car horn sounds.

TIGER. Yeah. Tell her, sorry for any trouble caused —

COLONEL. That's all right.

TIGER. Well, cheerio for now, then.

COLONEL. Cheerio, Tiger.

TIGER. I 'spect I'll be back soon.

COLONEL (*softly*). Forgive me.

TIGER (*softly*). 'Ere, chin up. "A Scout saying is Never Say Die 'til you're dead."

COLONEL. I'll remember.

TIGER. An' you will come to see me?

COLONEL. I promise.

TIGER. Right, off I goes then.

Cut to the campsite in the morning, with shouts of bathers in the background.

SPIDER. Master Miles! Time to get up!

   COOP *approaches.*

   Oh hello Coop. Just goin' for a bathe. Coming?

COOP. Not this morning.

SPIDER. Er — have you seen Tiger?

COOP. Isn't he bathing?

SPIDER. No — not in his tent either. Must've been up early to beat me. (*He calls.*) Come on Master Miles, all the others are in already! (*To* COOP.) I wondered, should we send out a search party?

COUP. Why? He's wandered off before.

SPIDER. Well, yes, but . . . Um — you and Tiger had words last night, didn't you?

COOP. After breakfast, get the lads on to something.

SPIDER. What?

COOP. Anything. Use your initiative.

SPIDER. A rope bridge across the stream I thought.

COOP. Fine. I'll be in my tent if you want me.

SPIDER. Aren't you coming to help?

COOP. Got one or two things to do. Carry on Spider.

COOP *goes*. MILES *approaches*.

SPIDER. Right (*He calls*.) Master Miles! Time for your bathe.

MILES. Oh my God.

Cut to the COLONEL, *wandering about the hall, distracted.*

COLONEL. Timms! Timms! Where the devil are you?

YOUNG GEORGINA. Daddy, Daddy, please! Tiger's not here, Daddy.

COLONEL. Timms, my suit was not pressed this morning, there was no breakfast!

YOUNG GEORGINA. You sent him away, Daddy.

YOUNG MILES. Georgie, I'm frightened.

COLONEL. That is a wicked lie, Georgina. Go to your room.

YOUNG GEORGINA. It's all right, Miles, hold my hand.

COLONEL. I know he's here. Timms! You lazy good for nothing! I'll find you. And when I do I'll tan your hide!

YOUNG GEORGINA. }
YOUNG MILES.   } Daddy!

Cut back to the present. The hall. RUFUS *growls as the front door latch goes and the door swings open.*

TIGER. 'Ello? 'Ello, anyone at 'ome?

The door shuts. RUFUS *welcomes him.*

'Ello boy. 'Ello Hereward. No, you're . . . Master at 'ome?

He walks on a little, his feet on the marble.

Hello! Only I just called to bring the clothes back. The butler's rig. Sorry it ain't pressed, only . . .

COLONEL (*very very distant*). Timms!

TIGER. Miss Georgina? Master Miles? Thankin' you all the same, but I don't think I shall be needin' it.

COLONEL (*as above*). Timms, where the devil are you?

TIGER. Uhh. Sorry, this buzzin' . . . sometimes . . . when I . . . I should've returned it before. But I bin away. And then Coop says the uniform don't . . . fit me.

COLONEL (*slightly closer*). I know you're here.

TIGER. You'd a liked Coop. 'E looks arter me. Only, I don't want to upset 'im. I'll just leave it at the foot of the . . . stairs. The old penguin suit (*He walks.*) There.

Pause.

(*Very softly, more sad than accusatory:*) You never come, did you? You said you would, but you never come to see me.

COLONEL (*closer; he remains slightly off for the rest of the scene, speaking 'in parallel' with* TIGER, *not quite 'to' him.*) I'd signed you see, I'd signed the paper.

TIGER. I've never said nothin' to Coop nor no one, and I know things wasn't easy for you. But — well — you said.

COLONEL. Tear it up, I said, it's a mistake, I signed in anger.

TIGER. I used to sit by the window every day. I couldn't understand it.

COLONEL. Can't tear up an official document, they said.

TIGER. I used to ask Coop. 'Excuse me, 'as there bin a message for me? I'm expectin' a visitor.' But there never was a message. I says to Coop, 'Why don't 'e come?' 'You set fire to his 'ouse, 'is property.' 'Yeah, but not all of it,' I says.

COLONEL. And after all, no one got hurt.

TIGER. '*That* don't matter,' says Coop, 'but an Englishman's property . . .'

COLONEL. Can I see him then? Visit him?

TIGER. They used to lock me in this room some days. And you know 'ow I 'ates bein' alone.

COLONEL. Not advisable, they said. You'll upset him. But I gave my word!

TIGER. And then . . . them things . . . them things they clipped on my 'ead . . . What for did they do that, eh? What for?

COLONEL. I know what goes on in those places.

TIGER. I says to 'em, 'Have you got permission?' 'Permission?' they says. 'From 'im, from the colonel?' And they all smiled. 'Oh yes, the colonel said it's all right.' Used to break old Coop up when they done that. Well, nothing he could do.

COLONEL. He's a menace to society, they said. Yes, but only my society, not yours. I can handle him.

TIGER. After a bit, I calmed down. Couldn't remember things too clear.

COLONEL. I blocked off the West Wing, and in time I wasn't entirely sure what had happened.

TIGER. I was sorry, what I done.

COLONEL. Still, we made a bit on the insurance. Kept us going. Then I'm afraid I rather took to what was left of the wine cellar.

TIGER. No drink down there. I'm TT now. Me!

COLONEL. Gradually, you see, things began to fall apart. My wife became unapproachable, the children ungovernable, and the estate — well, frankly, I ceased to care. Then came the day of the annual cricket match.

TIGER. We 'ad 'Leisure Activities' down there. You should've seen me makin' wicker lampshades! I'd say to Coop, 'What d'you think of that then?' And 'e'd look at it, then 'e'd look at me, and we'd both 'ave a good laugh.

COLONEL. You remember when the village sent up a team to play my Invitation XI in the park? You used to umpire with poor old Baldy. I can see you now, covered in sweaters, drunk as a lord. D'you recall the Lord Lieutenant chasing you round the boundary for giving him out on political grounds? You were good value.

TIGER. D'you know, we even 'ad an annual cricket match down there, just like those we 'ad in the park. Twenty-two of that lot jumpin' up and down. Laugh!

COLONEL (*with a little laugh*). I only scored seven that day. My worst performance for years. My heart wasn't in it.

TIGER. I joined in wiv some things but mostly I stayed on me own, goin' for walks an' that.

COLONEL. And walking back to the house afterwards I stopped in the clearing. The trees were showing the first tints of autumn. And I remembered the gypsies and telling the children that anyone could camp on common land, and the notice 'Hanged, drawn and quartered'. And I was overwhelmed with a sense of . . . desolation. I came up to the house and opened the door to the hall and called — no one.

TIGER. I used to imagine I 'eard you sometimes. 'Timms, where the devil are you?' And I'd just stand there 'til Coop come and got me.

COLONEL. I took off my jacket and tie. Orange and yellow MCC. And walked wearily up these stairs.

TIGER. 'Where is 'e?' I said to Coop. 'Summat's wrong. He'd 'ave come, else.'

COLONEL. And looking down it seemed the most natural thing in the world.

TIGER. "Never say die 'til you're dead."

COLONEL. I called once more. 'Timms!'

TIGER (*whispered with urgency*). No, you mustn't —

COLONEL. But no one answered, so —

TIGER. It don't matter about not comin', 'onest, I understands.

COLONEL. I slipped the tie back around my neck, closed my eyes and —

TIGER. NO!

RUFUS *barks*. TIGER *is very distressed.*

You silly sod, there's no need! . . . See 'im 'angin' there . . . See 'im kickin' and strugglin'. My mate . . . 'e 'ad no cause . . . Oh dear . . .

RUFUS *barks.*

GEORGINA (*from the top of the stairs*). Rufus! My head! What's the row about?

TIGER. 'E was a good man. What a waste!

GEORGINA. Why hello, Tiger.

TIGER. I never meant none of it really!

GEORGINA. Oh dear, are you having another little fit?

TIGER *blows his nose.*

TIGER. Sorry if I disturbed you, it's just when I looks up there I . . . (*He blows his nose.*)

GEORGINA. Something upsets you. Mmm. I see you've got your togs back.

TIGER. Master Miles, he brought 'em down.

GEORGINA. Did he indeed?

TIGER. So I'm returning' the old suit.

GEORGINA. No use to you, Tiger?

TIGER. I . . . liked bein' butler here, miss. But — things is different now.

GEORGINA. Yes . . . How about your young friend? The Boy David?

TIGER. Coop? I think you knows the answer to that.

GEORGINA. It could be altered to fit.

TIGER. You don't know 'im, miss.

GEORGINA. I'm so sorry if I was the cause of any friction between you.

TIGER (*beat*). I'll be getting back now. He'll be worryin' —

GEORGINA. Pity about the suit. I had thought that once you'd worn it again . . .

TIGER (*beat*). Miss Georgina, I don't owe you nothin'.

GEORGINA. No? It was from here, Tiger. Just about where I'm standing . . .

TIGER. The colonel, what 'e done, I'm as sorry as you are but I never made 'im.

GEORGINA. And now it seems you've added abduction to your talents.

TIGER. So I brought it back. He'd've understood.

GEORGINA. Never mind Daddy. What have you done with Miles?

TIGER. You know, I never said to 'im, but I'm glad 'e learned me the scouting. Given a lot o' pleasure to a lot o' lads.

GEORGINA. Timms —

TIGER. So, beggin' your pardon for the intrusion, but I think I'll be gettin' back now.

GEORGINA. *Where's my brother?*

Cut to campsite with sounds of laughing and splashing.

SPIDER. Come on Master Miles, jump in!

MILES. It looks jolly cold. Can't I just paddle?

SPIDER. Don't be a scaredy cat! Tucker, Pringle, come on, grab his arms!

TUCKER. Come on sir, it's lovely in.

PRINGLE. We'll all jump together.

MILES. But I might have forgotten how to swim!

SPIDER. You never forget. Ready? One, two, three!

ALL. Hurray!

   There is a big splash.

MILES. Oh, oh, oh! It's c-c-cold! Oh, oh, oh, gosh!

SPIDER. There, that wasn't so bad, was it? Oh, hello Tiger. You back?

TIGER (*approaching*). 'Lo, Spider. 'Ave you seen Coop?

SPIDER. In his tent I think. Everything all right?

TIGER. Never better, lad.

SPIDER. Good.

MILES. Can I g-get out n-now?

SPIDER. You haven't had a proper bathe 'til you've been right under.
Come on Scouts, duck him!

There are shouts and laughter, with MILES *yelling 'Pax, pax' etc.*
Cut to the tent. The flaps rustle as TIGER *enters.*

TIGER. Oh, 'ere you are. Ain't like you to be inside, lovely day like
today.

COOP. Where have you been?

TIGER. Bin for a walk. Gettin' worried was you?

COOP. Where d'you go?

TIGER. Up the 'ill, to the 'ouse.

COOP. Can't stay away, eh? Pass my sweater.

TIGER. Eh? Sweater? What's going on?

COOP. I'm packing.

TIGER. Packin'?

COOP (*beat*). I can't stay here.

TIGER. Well, praps you're right. Praps it would be difficult. Mind you,
the lads'll be disappointed. It's a wonderful campsite.

COOP. I don't mean all of us.

TIGER. Eh?

COOP. Just . . . me . . . go.

TIGER. On your own? Where?

COOP. I don't know. Anywhere.

TIGER (*sighs*). Aw, come on Coop! Don't be like that, not you. Look,
I'm sorry about last night. I lost me temper.

COOP. It's not just that.

TIGER. What then?

COOP (*beat*). I — didn't like to see you in those clothes.

TIGER. I know lad. That's why I took 'em back.

COOP. Eh?

TIGER. What d'you think I bin doin'? I ent as daft as I look you know.

A dog barks in the distance.

GEORGINA (*distant, outside*). Miles, where the hell are you?

TIGER. An' it's all over now, I promise. It'll just be you an' me an' the lads.

GEORGINA (*distant*). Miles!

TIGER. We'll spend the rest of the time doin' what you want. 'Ow about that?

The tent flap rustles.

PRINGLE. Sir, Tiger, sir!

TIGER. Only, don't go Coop, that's all.

COOP. What is it, Pringle?

PRINGLE. Sir, that lady's here.

TIGER. You're drippin' all over the tent.

PRINGLE. She wants to see you, sir.

TIGER. Does she? Right away lad . . .

COOP. That's it. Chop chop, run along.

TIGER (*beat*). It ain't been easy for me, Coop.

PRINGLE. She's got a gun, sir!

Cut to GEORGINA *by the stream; sounds of bathing.*

GEORGINA. Miles, will you stop arsing around and listen!

MILES. Can't hear old thing. Ears full of water.

GEORGINA. Where were you last night? I was worried sick.

MILES. You ought to come in, Georgie. It's lovely.

GEORGINA. You're to come home this instant! (*She stamps her foot.*) Miles!

TIGER (*quietly*) Praps . . . praps 'e don't want to, Miss Georgina.

GEORGINA. When I want *your* advice I'll ask for it.

TIGER. Praps he's enjoyin' 'is bathe.

GEORGINA. Miles, are you coming or must I blast you out of the water?

MILES. Wheee!

COOP. Looks to me as though he's having a decent time. For a change.

GEORGINA. Oh, the Boy David.

TIGER. Now you two —

GEORGINA. I understand you have declined my offer.

COOP. I would not put that suit on if I was standing naked on the Russian Steppes.

GEORGINA (*laughs*). Naked in Russia? How delicious, and how appropriate!

TIGER (*quietly firm*). All right Coop, I'll 'andle this.

COOP. Eh?

GEORGINA. Miles, are you intending to prat around in there all day while I'm being insulted?

MILES. But I'm rather enjoying being a Scout again.

GEORGINA. Then you can damn well enjoy it somewhere else. You're not camping here, any of you!

TIGER (*beat*). Beg pardon, miss?

GEORGINA. Words of one syllable, o toothless Tiger: get off my land!

Pause.

COOP (*with a contemptuous laugh*). Oh dear.

MILES. Hang on, I'm coming out.

GEORGINA (*to* TIGER). Did you hear? Off!

TIGER. But . . . this . . . ain't your land, Miss Georgina.

GEORGINA (*beat*). What?

TIGER (*quietly*). This is common land.

GEORGINA. Oh, the old egalitarian head rears itself again.

TIGER. He said, "I remember telling the children that anyone can camp on common land."

GEORGINA. You're an old man, your head is muddled.

TIGER. An' the gypsies always used to come here.

MILES (*approaching*). Now what seems to be the problem?

GEORGINA. Miles, you look ridiculous.

MILES. Sorry. It's the cold that makes it shrivel up.

TIGER (*quiet and calm*). So beggin' your pardon, the colonel says we can stay.

GEORGINA. I can handle a gun you know, Timms.

TIGER. This is not your land.

GEORGINA (*like a spoilt child*). 'Tis, it is, it is! Tell him, Miles!

MILES. You know he's right, Georgie. Now steady with the gun, old thing.

GEORGINA. But it's *next* to my land!

TIGER. And we've got a lot to do so if you'll excuse us —

GEORGINA. You're a wicked old man!

MILES. Hand it over, old girl. That's it. My prize possession the old Holland and Holland. Not for girls.

GEORGINA. You will come home won't you?

SPIDER. Oh, you going? What a pity.

MILES. Tell you what, when you're a patrol leader, you come and see us again eh?

SPIDER. Bob's your uncle!

MILES. You know you can stay here, Tiger, if you want to. And thanks for all the work you did yesterday.

TIGER. Our pleasure.

MILES. I'll get my things and we'll be off. Cheerio Tiger, it was good to see you again.

TIGER. And you Master Miles. Good luck.

MILES. Say goodbye, Georgie.

GEORGINA. Shan't.

MILES. We know what happens to little girls who sulk, don't we?

GEORGINA (*sulking*). Goodbye. Tiger.

TIGER. Goodbye, Miss Georgina.

MILES. Let's see if we can bag something for the pot on the way. Come on Rufus.

GEORGINA (*going*). I'll cook something nice for lunch.

MILES (*going*). And for pudding I want roly-poly. Lots and lots of . . . roly-poly.

GEORGINA. Of course, darling. As much as you want.

TIGER (*coughs*). Yeah well. Spider, get this lot dressed and breakfasted and we'll organise the day. Coop? (*He sighs.*) Oh dear.

Cut to the tent. Flaps rustle.

COOP. Have they gone?

TIGER. Yeah. Glad that's over.

COOP. I think they had the wrong one making wicker lampshades.

TIGER (*little laugh*). Now, now. (*Pause.*) Right, then —

COOP. Pass my mug and plate, will you?

TIGER. Oh, you still ain't on about that are you?

COOP. I . . . told you.

TIGER. But what about the campin', Coop? I got it all planned.
There'll be a wide game, and buildin' the bridge, and I shall need you
for makin' the rudimentary canoe.

COOP. And taking stones out of horses' hooves?

TIGER (*slight pause*). Grown up, 'ave you?

COOP. I just want to think things out.

TIGER. I never knows what people mean by that. Don't think they do,
either.

COOP. You can do without me. Spider's ready to take over.

TIGER. Do wivout you . . . It wasn't enough, was it? What I done out
there just now. What an old fool.

COOP. I wouldn't look up to an old fool.

TIGER. Dangerous to 'ave 'eroes Coop. 'Less they're in picture books.

COOP. Sorry to leave you in the lurch.

TIGER. Oh, I'm sure we'll manage, thank you. Spider's a good Scout,
an' he's got a missin' ingredient: keenness.

COOP (*quietly*). Tiger.

TIGER. 'Course, he ain't got your world weary experience that benefits
us all so greatly . . . (*Beat.*) Sorry, lad.

COOP. Well, I think that's the lot.

TIGER. Made your mind up 'ave you?

COOP. Yes.

TIGER (*smiles*). Got your matches? Useful things, matches.

COOP. I've got them.

The tent flap rustles.

SPIDER. Ah, here you are.

TIGER. What is it?

SPIDER. It's about the bridge. We're not going to have enough rope —

TIGER (*dully*). "We had no rope with us in West Africa so we used strong creeping plants."

SPIDER. Yes, but which ones?

TIGER. "Willow and hazel make good withes; you cannot tie knots with them but you can generally make a timber hitch."

SPIDER. Willow and hazel. There's a good lot of that around.

COOP. Spider, here, take this.

SPIDER. Eh? But it's your patrol leader flash.

COOP. You deserve it.

SPIDER. I don't understand.

TIGER. Coop is . . . leaving us. He's got to . . . go back.

SPIDER. Oh dear, what a pity. It won't be the same without you, Coop.

TIGER. No, it won't. Anyroad, you are now acting unpaid probationary patrol leader, Spider. Think you can do it?

SPIDER. Gosh, I don't know. A lot to live up to. I'll do my best.

COOP. Good luck, Spider.

SPIDER. See you back at base, then?

COOP. Yeah.

SPIDER. The rope we have got, it's all different thicknesses —

COOP. Use a sheet bend.

SPIDER. Not a reef?

COOP. Sheet bend's better.

SPIDER. Bob's your uncle.

SPIDER *goes.*

COOP. He won't let you down.

TIGER. Got everything?

COOP. Think so.

TIGER. Can I walk up the lane wiv you?

COOP. Course you can.

TIGER. Come on then.

Cut acoustic.

TIGER (*internal*). "Camp Fire Yarn Number Six. Tying Knots. Every Scout ought to be able to tie simple knots. To tie a knot seems an easy thing and yet there are right ways and wrong ways of doing it. A life may depend on a knot being properly tied."

COOP (*internal*). "The right kind of knot is one which you can be certain will hold under any amount of strain, and which you can undo easily if you wish to."

Fade up summer acoustic.

TIGER. Lovely day, innit? Smell that air.

COOP. God's air.

TIGER. Well, there she is. The old Bedford.

COOP. Think you'll be able to manage her all right?

TIGER (*gently*). I ain't lost all my faculties you know.

COOP. I'm sorry.

TIGER. Off you go then. Down the old A1.

COOP (*smiles*). The old A1.

TIGER. Coop, I just wants to say — you're a fine fella. A good Scout. It's been a pleasure and a privilege.

COOP. Thanks.

(*Internal.*) See him standing there, the sun glinting on his specs. Baggy shorts, cropped grey hair. Big smile on his face. An infinite capacity for forgiveness and keeping cheerful.

TIGER (*internal*). See him standin' there tryin' to sort it all out. Like lookin' in a forty-year-old mirror. Good-hearted boy. The sort o' lad a man could . . . wish for.

Cut to a hospital ward, not too busy.

COOP. Hello.

TIGER. 'Oo are you?

COOP. My name's Cooper. (*Beat.*) What's your name?

TIGER. Timms.

COOP. How d'you do, Mr Timms.

TIGER. Ow do.

COOP. Just arrived, haven't you?

TIGER. Not long. You stayin' 'ere too?

COOP. No, I visit. Few days a week.

TIGER. I'm not permanent neither. Just a rest like.

COOP. I'm here to help you, if you need anything. Or if you just want to talk.

TIGER. No thanks. I'm expectin' a visitor of me own.

COOP. Oh?

TIGER. The colonel, he'll be comin'.

COOP. I see.

TIGER. Thanks all the same.

COOP (*beat*). What's that you're reading?

TIGER. It's a book I bin lent.

COOP. *Scouting for Boys*. That's Baden-Powell, isn't it?

TIGER. Somethin' to read.

COOP. There seems to be a lot crossed out.

TIGER. Yeah. I dunno 'oo done that. Pity.

COOP. Are you a Scout?

TIGER. Manner of speakin'. You?

COOP. I don't know much about it. I'd like to hear.

TIGER. Would you? Honest?

COOP. Will you tell me?

TIGER. All right, Mr Cooper.

COOP. Most people call me Coop.

TIGER. I'm Tiger to me friends.

COOP. Hello, Tiger.

TIGER. 'Ello, Coop.

Fade out into summer acoustic.

TIGER. Before you go. 'Ere. Summat for you.

COOP. What?

TIGER. Go on, take it.

COOP. But it's your copy. The one he gave you.

TIGER. Like you to 'ave it.

COOP. Thanks.

TIGER. I wrote in it.

COOP (*pages rustle*). 'For me old china Coop, with love from T. Timms.' Oh Tiger, Tiger . . .

TIGER. Still burnin' bright, lad, burnin' bright!

COOP. Always burning bright. Well . . .

TIGER. Coop — it . . . it wasn't all bad, was it?

COOP. Bad? I wouldn't have missed it for the world.

TIGER. Really? That's nice.

COOP. Look Tiger, I —

TIGER. No, lad. You made your decision. Stick to it.

COOP. I'll take care of the book.

TIGER. And yourself lad. Remember "Once a Scout —

COOP. — Always a Scout." Yeah.

TIGER. I ain't much good at advice an' that, but well — you got somethin' Coop. Ain't given to everyone. 'E 'ad it, the old colonel. Praps I did, I dunno. Even Master Miles 'as, in a way. Miss Georgina, maybe not. But you 'as. It's a sort of touch. You 'ang on to it, lad. I know you gets angry, but, well, don't throw it away.

COOP. I'll remember.

TIGER. And you knows where to find us. You'll always be welcome.

COOP. I know.

TIGER. Yeah, well, off you go.

COOP. Cheerio, Tiger.

TIGER. Cheerio then, Coop.

Fade in 'Ging-Gang' quietly.

TIGER (*internal*). "I hope I have been able in this little book to show you something of the appeal that lies in scouting for all of us. This brotherhood of scouting is in many respects similar to a crusade. Besides the adventure to be gained, Scouts from all parts of the world are ambassadors of good will, making friends, breaking down barriers of colour or creed and of class. That surely is a great crusade."

COOP (*off*). So long, Tiger.

TIGER (*calls*). Mind 'ow you go lad.

(*Internal.*) "Scouting is a fine game if we put our backs into it and tackle it well. Play up! Each man in his place and play the game. There are great times ahead — we shall need you!"

Pause.

(*Softly.*) I . . . was a boy . . . once.

Song fades up and out.

The end.